DEPARTMENT OF THE ARMY FIELD MANUAL

SUBMACHINE GUNS CALIBER .45 M3 AND M3A1
FIELD MANUAL

By DEPARTMENT OF THE ARMY
JULY 1957

©2013 Periscope Film LLC
All Rights Reserved
ISBN#978-1-940453-11-8
www.PeriscopeFilm.com

DISCLAIMER:

This document is a reproduction of a text first published by the Department of the Army, Washington DC. All source material contained herein has been approved for public release and unlimited distribution by an agency of the US Government. Any US Government markings in this reproduction that indicate limited distribution or classified material have been superseded by downgrading instructions promulgated by an agency of the US government after the original publication of the document No US government agency is associated with the publication of this reproduction. This manual is sold for historic research purposes only, as an entertainment. It contains obsolete information and is not intended to be used as part of an actual training program. No book can substitute for proper training by an authorized instructor.

©2013 Periscope Film LLC
All Rights Reserved
ISBN#978-1-940453-11-8
www.PeriscopeFilm.com

*FM 23-41

FIELD MANUAL
NO. 23-41

HEADQUARTERS,
DEPARTMENT OF THE ARMY
WASHINGTON 25, D. C., *8 July 1957*

SUBMACHINE GUNS, CALIBER .45, M3 AND M3A1

		Paragraphs	Page
CHAPTER 1.	INTRODUCTION		
Section I.	General	1, 2	2
II.	Description	3-5	2
CHAPTER 2.	MECHANICAL TRAINING		
Section I.	Disassembly and assembly	6-15	7
II.	How the submachine gun functions	16-26	25
III.	Operation	27-32	29
IV.	Malfunctions, stoppages, and immediate action	33-36	32
V.	Care and cleaning	37-46	34
VI.	Repair parts and accessories	47, 48	40
VII.	Ammunition	49-56	40
CHAPTER 3.	MANUAL OF ARMS	57-62	44
4.	MARKSMANSHIP TRAINING		
Section I.	General	63-65	47
II.	Preparatory marksmanship training	66-76	47
III.	Courses fired	77-80	58
IV.	Range firing	81-83	60
V.	Targets, ranges, and range safety precautions	84-86	67
CHAPTER 5.	MARKSMANSHIP, MOVING GROUND TARGETS AND VEHICULAR FIRING		
Section I.	Firing at moving ground targets	87-91	72
II.	Moving target range and safety precautions	92-96	73
III.	Vehicular firing (open vehicle)	97-100	76
CHAPTER 6.	TECHNIQUE OF FIRE AND DESTRUCTION of MATERIEL		
Section I.	Technique of fire	101-103	80
II.	Destruction of materiel	104-108	82
CHAPTER 7.	ADVICE TO INSTRUCTORS		
Section I.	General	109-113	83
II.	Training schedule and training notes for submachine gun marksmanship course	114-117	84
III.	Training schedule and training notes for submachine gun familiarization course	118-120	86
IV.	Range firing	121-124	87
V.	Score card	125	88
VI.	Training aids	126-128	90
CHAPTER 8.	SAFETY PRECAUTIONS	129-132	96
APPENDIX.	REFERENCES		99
INDEX			100

*This manual supersedes FM 23-41, 10 August 1949, including C 1, 12 February 1952, and C 2, 6 January 1956.

TAGO 7261-B, July

1

CHAPTER 1

INTRODUCTION

Section I. GENERAL

1. Purpose and Scope

This manual is a guide for unit commanders in training their men in the use of the Submachine Guns, Caliber .45, M3 and M3A1. It explains how to disassemble, assemble, fire, and take care of the weapons. It describes the parts and explains how they work. The step-by-step arrangement of the text provides for progressive training, promotes learning, and aids in organizing and presenting instruction.

2. Importance of Submachine Gun Training

a. The submachine gun is the individual weapon of many men in the Army. It is issued as on-vehicle materiel with most combat vehicles, to furnish automatic fire at close ranges or when a crew must dismount from a disabled vehicle. The submachine gun in the hands of a well-trained soldier is a very effective combat weapon.

b. Learning to hit the target and to keep the weapon in excellent operating condition are the most important steps in submachine gun training.

Section II. DESCRIPTION

3. General Description

The Submachine Guns, Caliber .45, M3 and M3A1 (hereafter referred to as the M3 or the M3A1), are air-cooled, blow-back operated, magazine-fed, automatic shoulder weapons (figs. 1-4). They are light, compact, and rugged. The stock is one piece of formed steel rod which can be telescoped for ease of handling; its ends are drilled and tapped so that it can be used as a cleaning rod. The stock may also be used as a disassembly tool or wrench. The stock of the M3A1 has a hand loader which is used to load the magazine. There is no provision for semiautomatic fire; however, because of the low cyclic rate of fire, the firer can fire single shots by proper trigger manipulation. Both weapons are fed from a box-type magazine which has a capacity of 30 rounds.

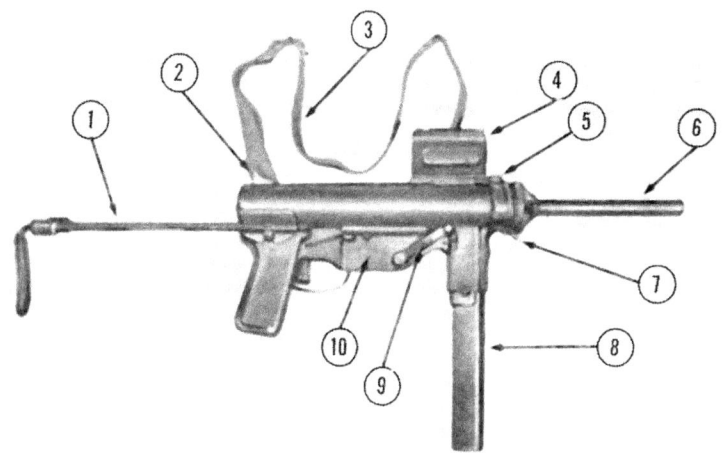

1 Stock
2 Rear sight
3 Sling
4 Cover
5 Front sight
6 Barrel
7 Barrel ratchet
8 Magazine
9 Retracting handle
10 Housing assembly

Figure 1. Submachine Gun, M3, right side, stock extended.

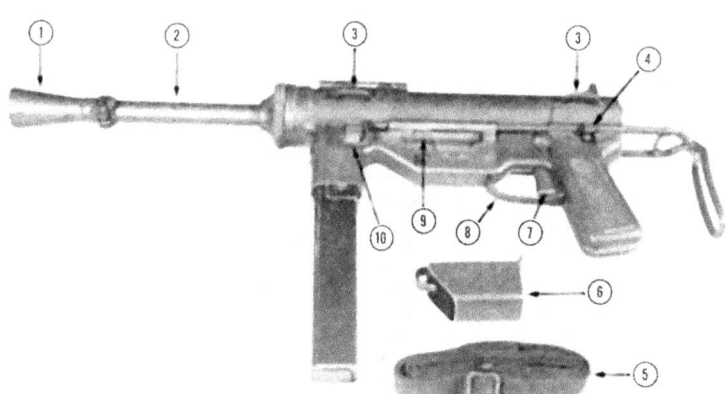

1 Flash hider (accessory)
2 Barrel
3 Sling loop
4 Stock catch
5 Sling
6 Magazine loader (accessory)
7 Trigger
8 Trigger guard
9 Oiler
10 Magazine catch

Figure 2. Submachine Gun, M3, left side, stock telescoped, sling removed, with flash hider.

TAGO 7261-B, July   3

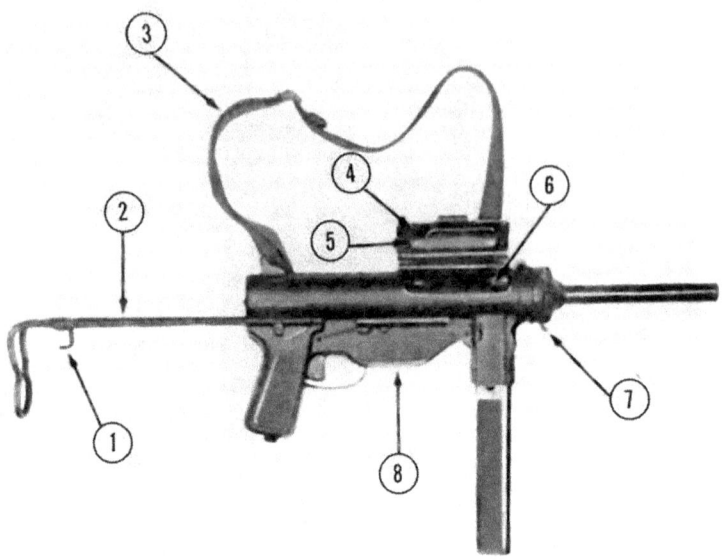

1 Hand loader and cleaning rod stop
2 Stock
3 Sling
4 Cover
5 Safety lock
6 Cocking slot
7 Barrel ratchet
8 Housing assembly

Figure 3. Submachine Gun, M3A1, right side, stock extended.

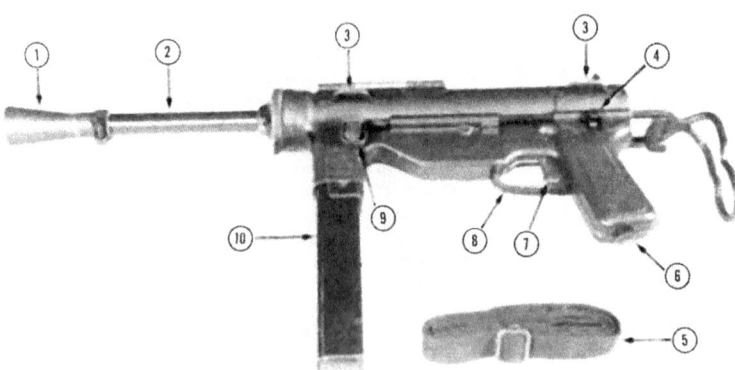

1 Flash hider (accessory)
2 Barrel
3 Sling loop
4 Stock catch
5 Sling
6 Oiler cap
7 Trigger
8 Trigger guard
9 Magazine catch
10 Magazine

Figure 4. Submachine Gun, M3A1, left side, stock telescoped, sling removed, with flash hider.

4. General Data

a. Barrel.

Diameter of bore	0.45 inch
Number of grooves	4
Twist in rifling	Uniform, right, one turn in 16 inches
Length of barrel	8 inches

b. Gun.

Length, over-all with stock extended	29.8 inches
Distance between sights	10 7/8 inches
Weight without magazine (approx)	8.15 pounds
Weight with 30 rounds in magazine (approx)	10.25 pounds
Weight of 30-round magazine (empty)	.75 pound
Weight of 30-round magazine (loaded)	2.10 pounds

c. Miscellaneous.

Chamber pressure (approx)	12,000 to 16,000 pounds per square inch
Muzzle velocity (approx)	900 feet per second
Cyclic rate of fire	450 rounds per minute
Sights	100 yards, fixed peep
Maximum range	1700 yards
Maximum effective range	100 yards
Trigger pull (approx)	5 to 7 pounds
Pull to cock weapon—M3	18 to 23 pounds
M3A1	10 to 12 pounds

5. Differences in Models

The M3 and the M3A1 are basically alike. However, when the M3A1 was developed from the M3 the following changes were made:

a. The retracting handle assembly, retracting lever assembly, retracting lever spring, and oiler clip have been eliminated.

b. A cocking slot has been cut into the top front portion of the bolt, so that the firer can retract the bolt with his finger. There is an ejector groove on the bottom of the bolt, extending the entire length of the bolt, to permit removal of the bolt and guide rod group without removing the housing assembly. The retracting pawl notch has been eliminated, and a clearance slot for the cover hinge rivets has been added.

c. The ejection opening and the cover assembly are longer. This allows the bolt to be drawn back far enough to be engaged by the sear. The safety lock is located farther to the rear on the cover.

d. An oil reservoir and oiler have been placed in the pistol grip of the receiver assembly. The stylus on the oiler cap may be used as a drift to remove the extractor pin. The barrel ratchet has been redesigned to provide a longer depressing level for easier disengagement from the barrel collar.

e. A bracket has been welded at the rear end of the stock. This bracket is used as a hand loader for loading ammunition into the magazine; it also serves as a cleaning rod stop.

f. The barrel collar has two flat cuts to permit the use of the stock as a wrench to unscrew a tight barrel assembly.

CHAPTER 2

MECHANICAL TRAINING

Section I. DISASSEMBLY AND ASSEMBLY

6. General

The submachine gun will function correctly if it is kept clean and is properly oiled and cared for. This chapter explains disassembly, assembly, functioning, care and cleaning, stoppages, and immediate action. It is a guide for mechanical training and outlines the procedures to be followed.

7. Nomenclature

The names of the parts of the submachine gun (figs. 5–10) should be learned during instruction in disassembly and assembly. As the instructor names the parts, the soldier repeats them to himself; he then names each part as it is removed and as it is replaced. Generally, the parts are named for the jobs they do. For example, the trigger guard actually guards the trigger, so that some object will not accidentally brush against the trigger and trip it.

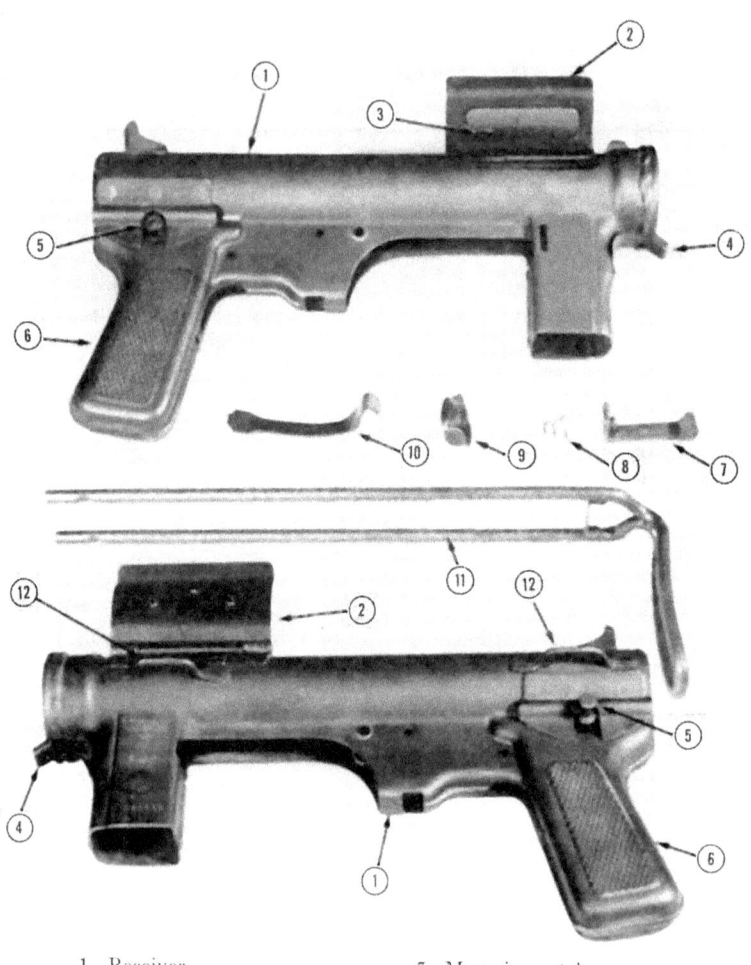

1 Receiver
2 Cover
3 Safety lock
4 Barrel ratchet
5 Stock catch
6 Pistol grip
7 Magazine catch
8 Magazine catch spring
9 Magazine catch shield
10 Trigger guard
11 Stock
12 Sling loop

Figure 5. M3 receiver, stock, and magazine catch assembly.

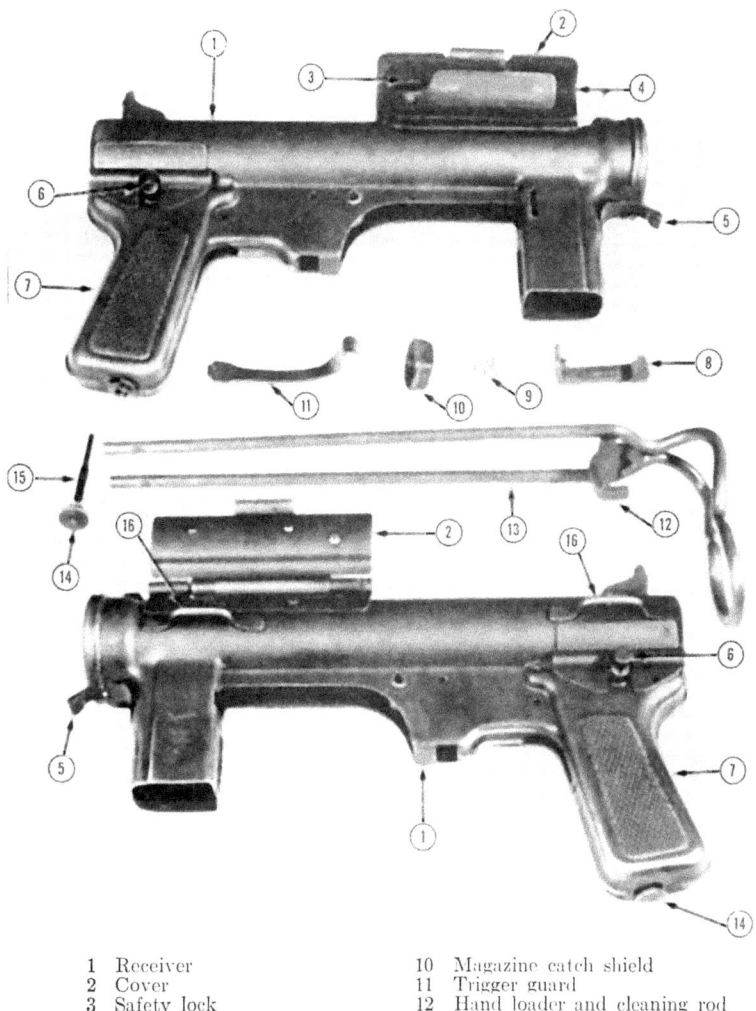

1 Receiver
2 Cover
3 Safety lock
4 Cover seal
5 Barrel ratchet
6 Stock catch
7 Pistol grip
8 Magazine catch
9 Magazine catch spring
10 Magazine catch shield
11 Trigger guard
12 Hand loader and cleaning rod stop
13 Stock
14 Oiler cap
15 Stylus
16 Sling loop

Figure 6. M3A1 receiver, stock, and magazine catch assembly.

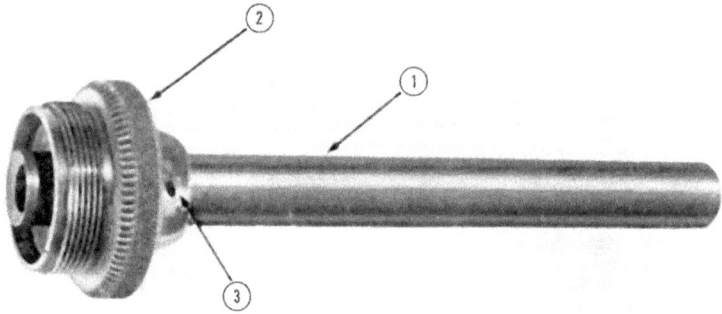

1 Barrel 2 Barrel collar 3 Barrel collar retaining pin

Figure 7. M3 barrel assembly.

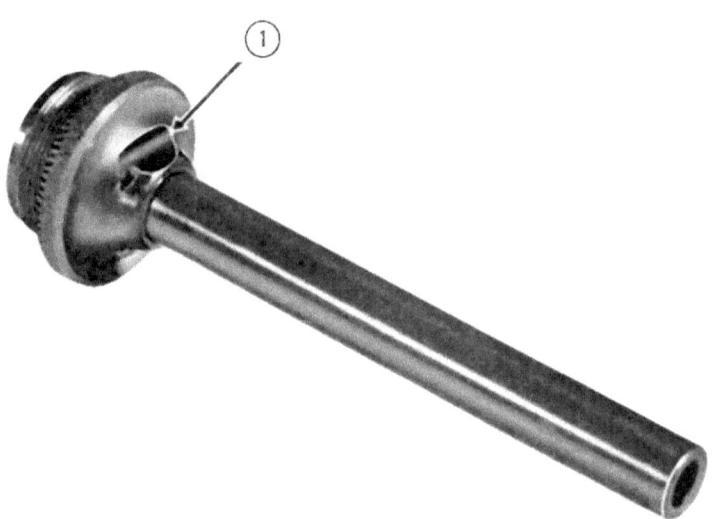

1 Disassembly cut

Figure 8. M3A1 barrel assembly.

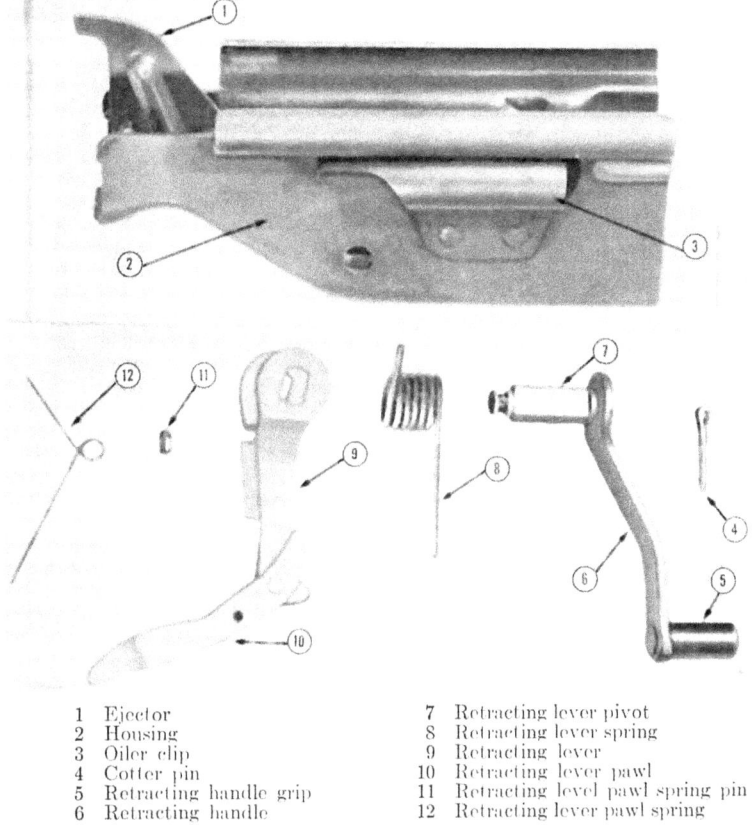

1	Ejector	7	Retracting lever pivot
2	Housing	8	Retracting lever spring
3	Oiler clip	9	Retracting lever
4	Cotter pin	10	Retracting lever pawl
5	Retracting handle grip	11	Retracting level pawl spring pin
6	Retracting handle	12	Retracting lever pawl spring

Figure 9. M3 housing assembly.

8. Disassembly, General

a. The soldier is permitted to disassemble only certain parts of the submachine gun—not because he cannot learn to disassemble all of them, but because unnecessary disassembly causes extra wear. Also, disassembly and assembly of some parts require special tools that are not normally available in troop units.

b. The left-hand column on the following chart shows those parts that may be disassembled by the soldier. The center column indicates those parts that the unit armorer may remove, including the parts disassembled by the soldier. The right-hand column shows those parts that only ordnance personnel may disassemble.

1 Ejector

Figure 10. M3A1 housing assembly.

Disassembly authorized	Disassembly performed by		
	Individual soldier	Unit armorer	Ordnance personnel
Field disassembly	X	X	
Barrel assembly			X
Housing assembly			X
Trigger and sear group	X	X	
Bolt and guide rod group	X	X	
Extractor		X	
Magazine	X	X	
Receiver			X

9. Guides to Follow in Disassembly and Assembly

These guides should be followed when disassembling and assembling the submachine gun.

a. Follow the step-by-step explanation in disassembling the submachine gun.

b. Do not attempt to disassemble or assemble the weapon against time.

c. If it is necessary to apply force, do it carefully so that none of the parts are damaged.

d. As the weapon is disassembled, line up the parts in the order of their removal. This procedures helps in assembly of the weapon, which is done in reverse order of disassembly.

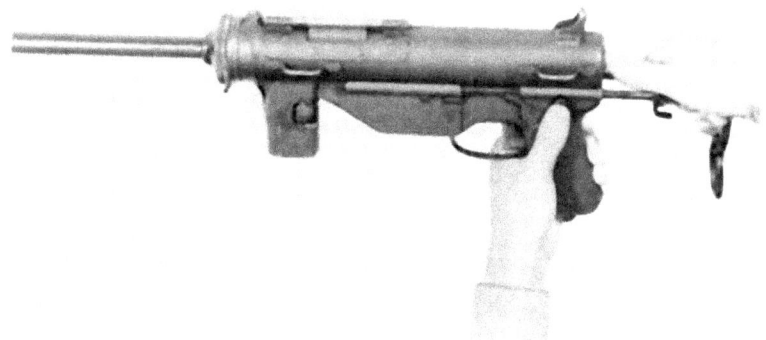

Figure 11. Press in on the stock catch on the left side of the pistol grip, and remove the stock by pulling it directly to the rear.

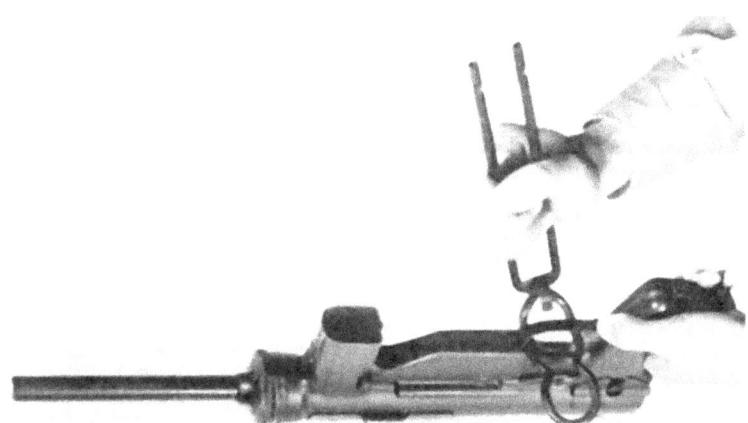

Figure 12. To remove the trigger guard, place one side of the shoulder rest of the stock on the housing assembly, against the trigger guard, and pry the trigger guard out of the pistol grip. Rotate the trigger guard toward the front of the weapon, and unhook the trigger guard from the housing assembly.

10. Field Disassembly

The soldier must learn field disassembly (removal of the groups) so well that he can perform this operation in the dark. The submachine gun can be field disassembled without special tools. Parts of the weapon are used instead of tools.

11. Procedure for Field Disassembly

a. Before disassembling the submachine gun, make sure that the weapon is *clear*. Press in on the magazine catch, and remove the

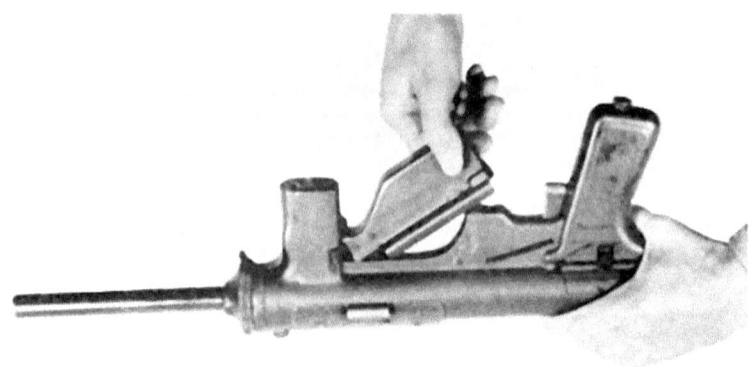

Figure 13. Remove the housing assembly by pulling up and to the rear on the rear end.

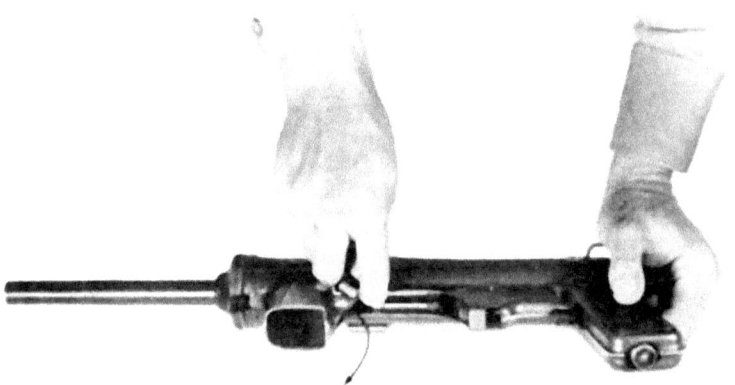

Figure 14. Remove the magazine catch assembly by rotating it toward the right side of the receiver.

magazine. Raise the cover, pull the bolt to the rear, and inspect the chamber. Allow the bolt to go forward by squeezing the trigger. Close the cover.

Note. To pull back (retract) the bolt on the M3, pull the retracting handle to the rear. To pull back the bolt on the M3A1, insert a finger into the cocking slot on the bolt and pull the bolt to the rear.

b. To disassemble the gun, follow the procedure illustrated in figures 11 through 21.

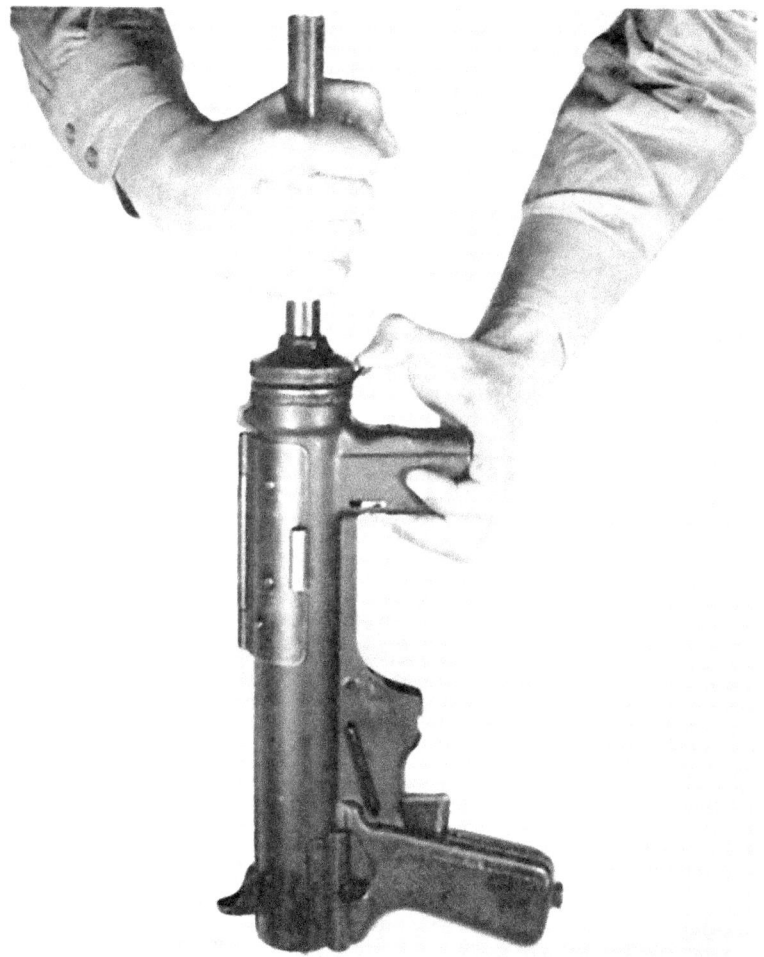

Figure 15. To remove the barrel, make sure the bolt is forward, depress the barrel ratchet, and unscrew the barrel. Do not allow the barrel ratchet to contact the notches in the barrel collar when removing or replacing the barrel.

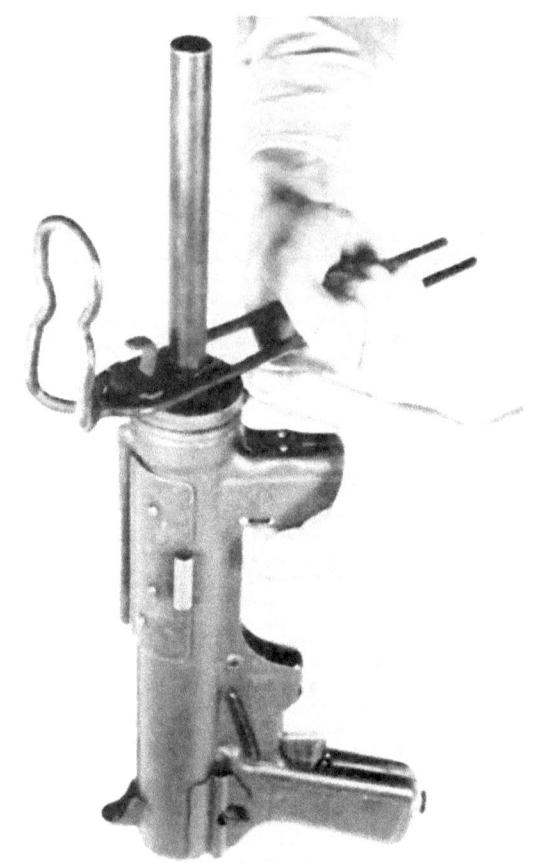

Figure 16. Using M3A1 stock as a wrench.

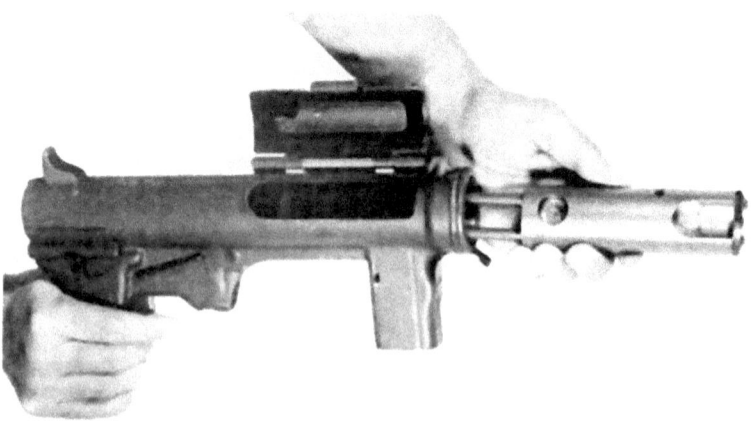

Figure 17. Open the cover, and withdraw the bolt and guide rod group from the receiver.

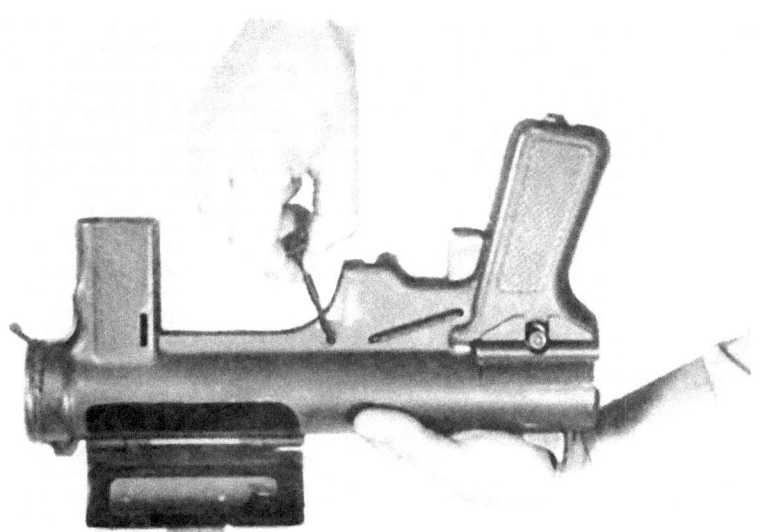

Figure 18. Drift out the sear pin. The magazine catch, ejector, or oiler stylus may be used as a drift.

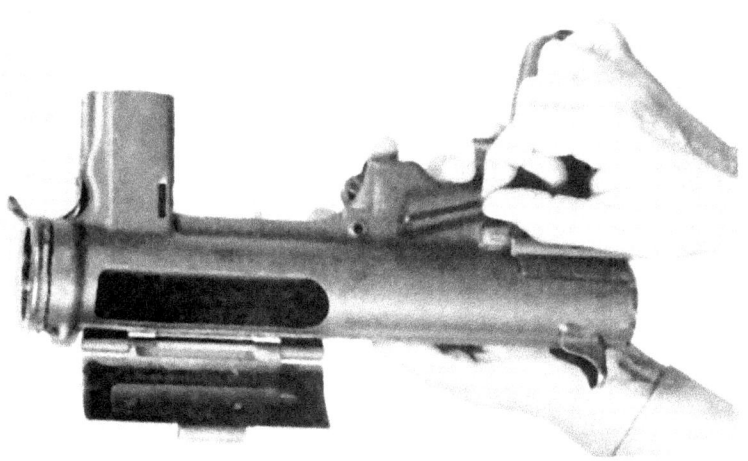

Figure 19. Remove the trigger pin.

Figure 20. Withdraw the trigger and sear group from the receiver. Be careful not to drop the connector pin. This completes field disassembly.

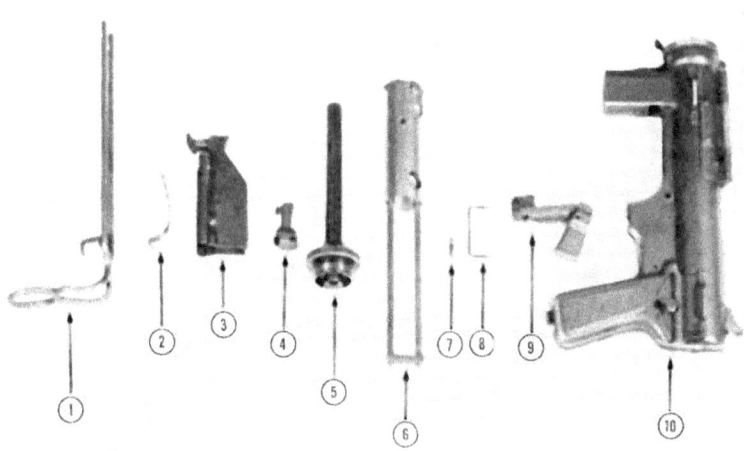

1 Stock
2 Trigger guard
3 Housing assembly
4 Magazine catch assembly
5 Barrel
6 Bolt and guide group
7 Sear pin
8 Trigger pin
9 Trigger and sear group
10 Receiver

Figure 21. Parts lined up in order of removal during field disassembly.

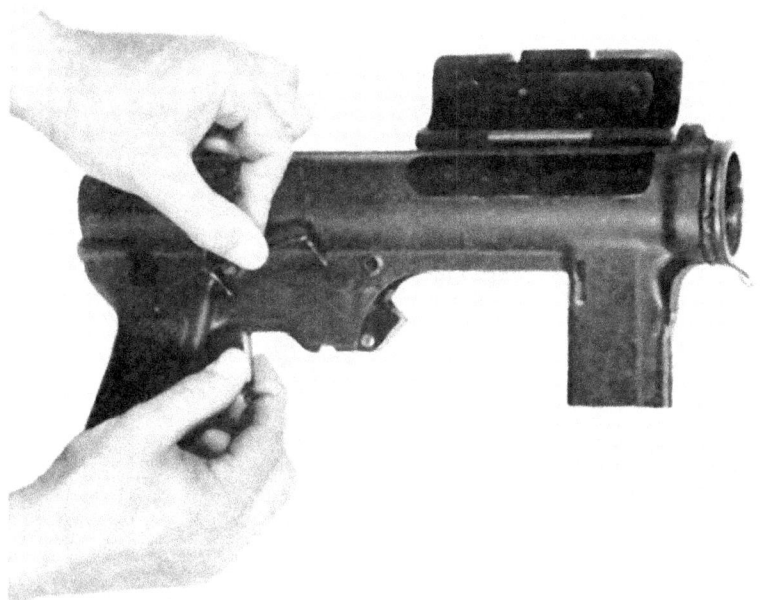

Figure 22. Replace the trigger and sear group in the receiver, with the trigger pin holes in the trigger and receiver alined. Replace the trigger pin. The front arm of the trigger pin goes through the holes in the receiver and does not go through any other parts.

12. Procedure for Assembly After Field Disassembly

a. The first steps in assembling the submachine gun are illustrated in figures 22 and 23.

b. Replace the bolt and guide rod group into the receiver with the retaining plate to the rear and the sear notch down. Close the cover. Depress the barrel ratchet, and replace the barrel by screwing the barrel collar all the way down until it is snug against the receiver. Replace the magazine catch assembly. Place the front projection on the housing assembly into its recess in the magazine guide. Press the rear end of the housing into place; make certain that it is properly seated. Insert the forward end of the trigger guard in its slot in the housing assembly, and rotate it to the rear (fig. 24). Do not use force in this operation. Press the rear end of the trigger guard until it snaps into its slot in the pistol grip. Press in on the stock catch, and replace the stock.

13. Disassembly of Groups (Detailed Disassembly)

a. *Magazine* (fig. 25). Lift the tab in the base plate by inserting a screwdriver in the hole. Remove the base plate, placing the fingers

Figure 23. Grasp the receiver, with the little finger on the trigger and the first finger against the sear. Press down on the trigger, and at the same time manipulate the sear until the sear pin hole in the sear is alined with the hole in the receiver. Replace the sear pin.

over the bottom of the magazine to prevent the magazine spring from flying out. Remove the magazine spring and the magazine follower.

b. Bolt and Guide Rod Group (figs. 26–30). Compress the driving springs, and remove the guide rod retaining clip. Remove the guide rod locating plate. Remove the bolt and driving springs from the guide rods. Do not remove the guide rod retaining plate from the two guide rods (this is done by ordnance personnel only). The extractor is removed for replacement only (this should be done by the unit armorer). To remove the extractor, drift out the extractor pin (from the small end, located on the bottom of the bolt). Place the rim of a dummy cartridge under the lip of the extractor, and lift it out.

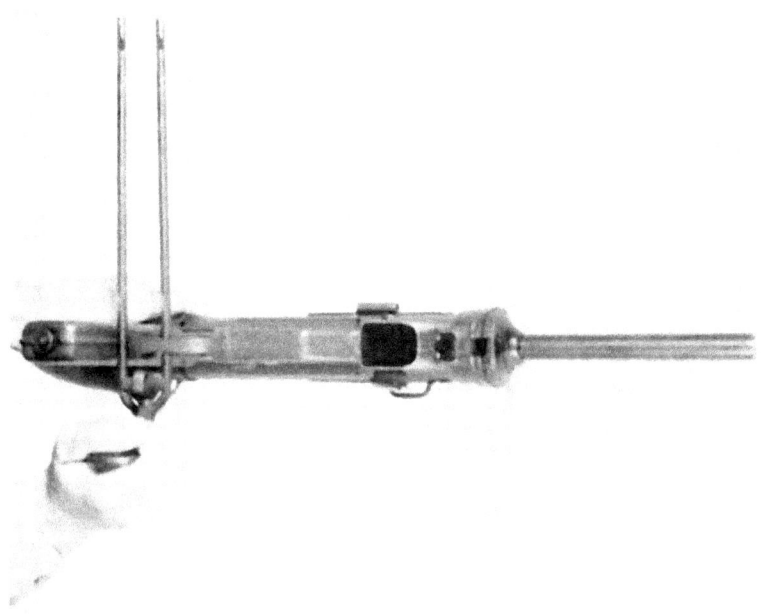

Figure 24. Use of the stock in replacing the trigger guard.

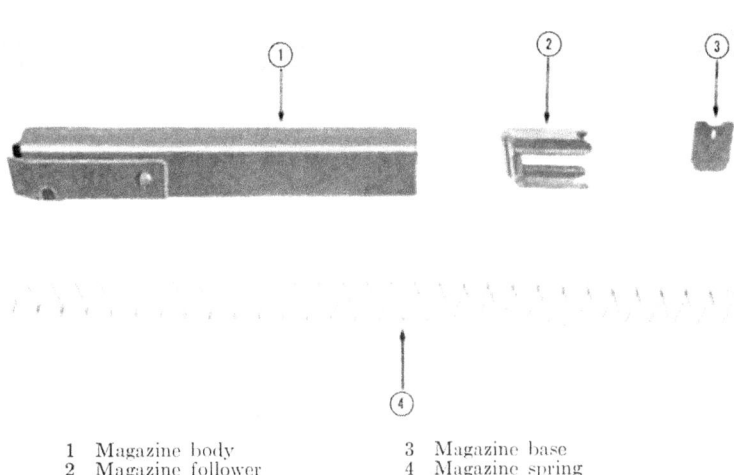

1 Magazine body
2 Magazine follower
3 Magazine base
4 Magazine spring

Figure 25. Magazine assembly.

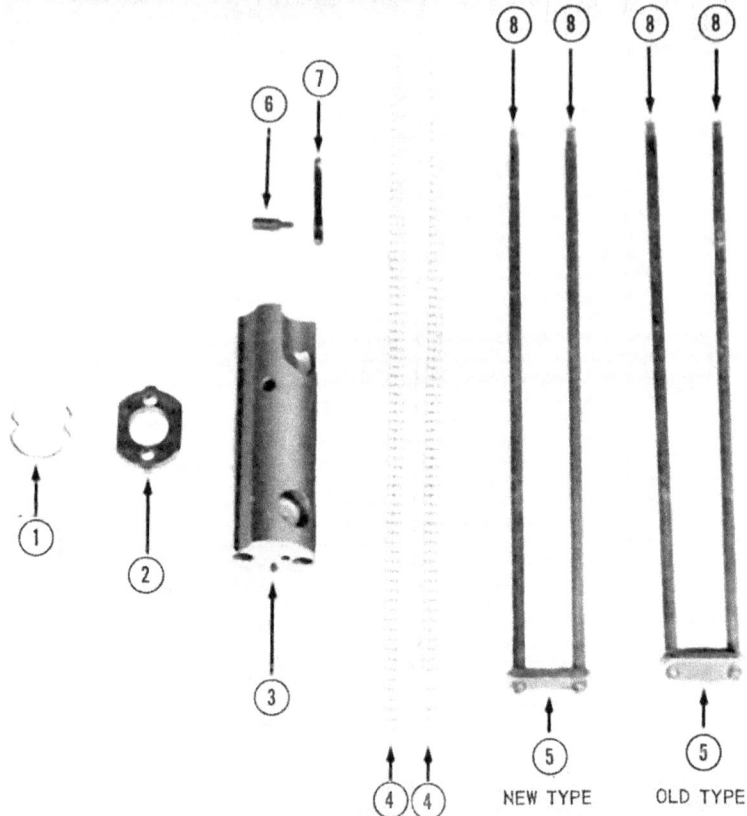

1 Guide rod retaining clip
2 Guide rod locating plate
3 Bolt
4 Driving springs
5 Guide rod retaining plate
6 Extractor pin
7 Extractor
8 Guide rods

Figure 26. Bolt and guide rod group.

c. *Trigger and Sear Group* (fig. 31). Drift out the connector pin. Remove the sear. Unfasten the trigger spring from the connector (do not remove it from the trigger).

d. *Housing Assembly.* The housing assembly is disassembled only by ordnance personnel.

14. Assembly of the Groups

a. *Magazine.* Replace the magazine follower and the magazine spring, with the loop toward the front of the magazine. Compress the magazine spring into the magazine body, and replace the magazine base. Press the tab back into its original position.

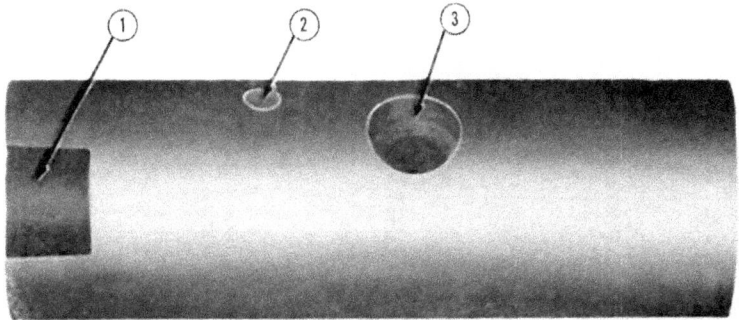

1 Clearance cut to allow ejection of spent cartridge case
2 Extractor pin
3 Safety lock recess

Figure 27. Top view of M3 bolt.

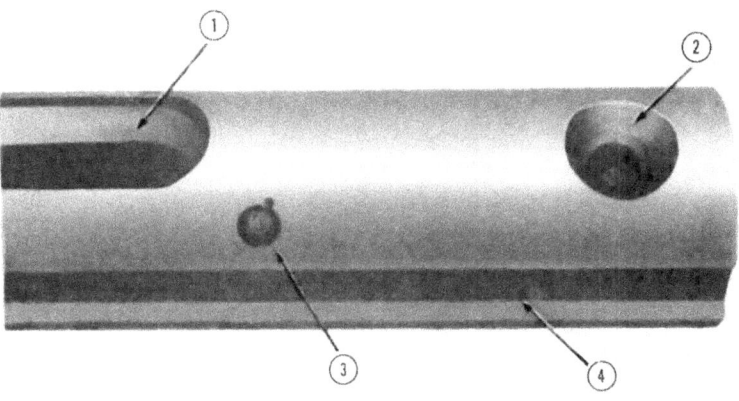

1 Cocking slot
2 Safety lock recess
3 Extractor pin
4 Clearance groove for cover rivets

Figure 28. Top view of M3A1 bolt.

b. Bolt and Guide Rod Group. Replace the extractor so that the notch and hole for the extractor pin in the bolt are alined. Firmly seat the extractor pin, and stake it in place. Place the driving springs on the guide rods. Compress the driving springs, and place the bolt on the guide rods with the firing pin away from the guide rod retaining plate. Replace the guide rod locating plate and the guide rod retaining clip.

c. Trigger and Sear Group. Fasten the trigger spring to the connector. Place the sear on the connector, with the sear nose up and to the rear. Replace the connector pin.

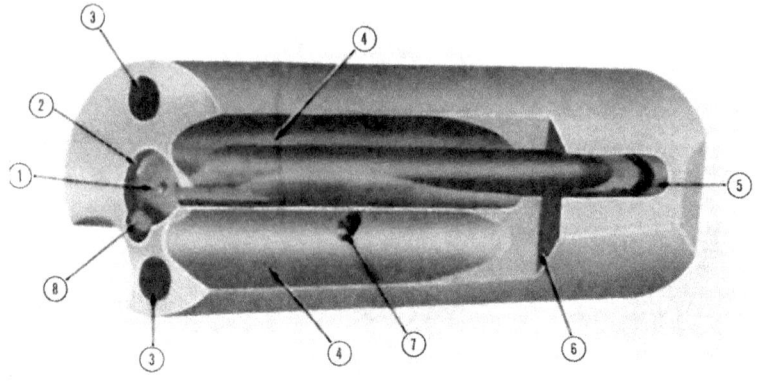

1 Firing pin
2 Cartridge recess
3 Guide rod bearing
4 Magazine lip recess
5 Retracting lever pawl notch
6 Sear notch
7 Extractor pin
8 Extractor

Figure 29. Bottom view of M3 bolt.

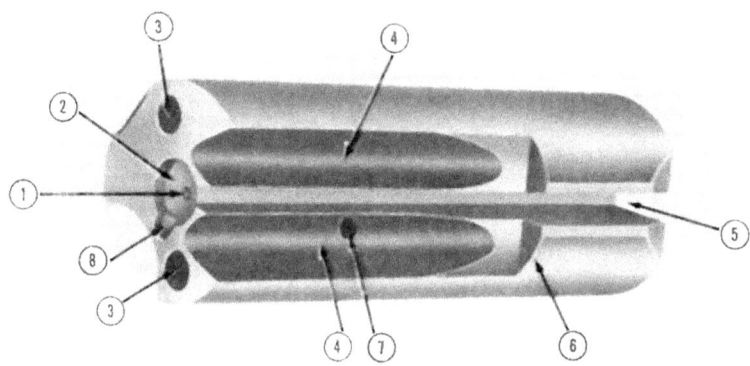

1 Firing pin
2 Cartridge recess
3 Guide rod bearings
4 Magazine lip recess
5 Ejector groove
6 Sear notch
7 Extractor pin
8 Extractor

Figure 30. Bottom view of M3A1 bolt.

15. Operation Check

After the weapon has been assembled, it should be checked to insure that it has been correctly assembled.

a. Pull the bolt to the rear sharply. It should be engaged and held to the rear by the sear.

b. Close the cover, and squeeze the trigger. The bolt should not move forward.

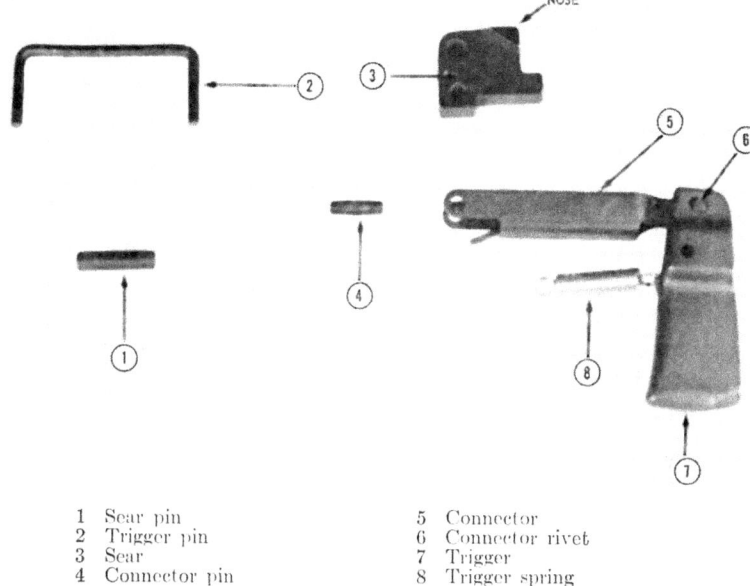

1 Sear pin
2 Trigger pin
3 Sear
4 Connector pin
5 Connector
6 Connector rivet
7 Trigger
8 Trigger spring

Figure 31. Trigger and sear group.

c. Open the cover, and squeeze the trigger. The bolt should move forward.

d. With the trigger held to the rear, pull the bolt to the rear and release it. The bolt should not be held to the rear by the sear, but should move forward.

Section II. HOW THE SUBMACHINE GUN FUNCTIONS

16. General

a. By disassembling and assembling the submachine gun, the soldier becomes familiar with the parts. The next step is to learn how these parts function. If the soldier understands how the submachine gun works, he will be able to keep it in operating condition and reduce any stoppage which might occur during firing. This knowledge will give the soldier confidence in his weapon.

b. Each time a cartridge is fired, the parts inside the submachine gun function in a given order. This is known as the cycle of operation (functioning).

c. The cycle of operation of small arms is broken down into eight basic steps. However, in the submachine gun, two of these steps—locking and unlocking—do not occur. The six basic steps of the cycle of operation of the submachine gun are listed below in the proper sequence, although more than one step may be occurring at the same time.

  (1) Feeding—the placing of a cartridge in the receiver, in front of the bolt, so it can be chambered. This action takes place in the magazine only.
  (2) Chambering—moving the cartridge forward until it is properly seated in the chamber.
  (3) Firing—the striking of the primer of the cartridge by the firing pin to ignite the cartridge.
  (4) Extraction—removal of the empty cartridge case from the chamber.
  (5) Ejection—removal of the empty cartridge case from the receiver.
  (6) Cocking—retraction of the bolt far enough so that it will pick up a new cartridge and, as it moves forward, will have enough energy to fire the new cartridge.

17. Functioning of the Submachine Gun, General

As the bolt is moved back to a cocked position, the driving springs are compressed, and the sear engages the sear notch of the bolt. When the trigger is pressed, the sear releases the bolt, which is driven forward by the driving springs. During this forward movement, the bolt pushes a cartridge from the magazine into the chamber. The bolt continues forward and fires the cartridge. When the cartridge is fired, the chamber pressure forces the bullet out of the muzzle of the barrel. At the same time, this pressure overcomes the forward movement of the bolt and starts it to the rear. By the time the bolt and empty cartridge case have moved to the rear far enough to open the rear end of the chamber, the bullet has left the barrel, and the chamber pressure has decreased. (In the submachine gun, the chamber pressure is relatively low and the bolt is relatively heavy; this eliminates the need for the steps of locking and unlocking.) During the rearward movement of the bolt, the empty cartridge case is extracted and ejected, the driving springs are compressed, and the top round in the magazine moves up against the lips of the magazine (fig. 32). The rearward movement of the bolt is stopped by the compressed driving springs or when it contacts the guide rod retaining plate.

18. Operation of the Trigger and Sear Group

a. When the trigger is pressed, it rotates around the rear arm of the trigger pin and forces the connector forward. This rotates the sear around the sear pin, causing the sear nose to be moved down and away from the sear notch in the bottom of the bolt. This allows the bolt to move forward under the action of the expanding driving springs.

b. If the trigger is held to the rear, the nose of the sear cannot engage the sear notch. The bolt will continue to move forward and backward, firing the weapon automatically until the trigger is released.

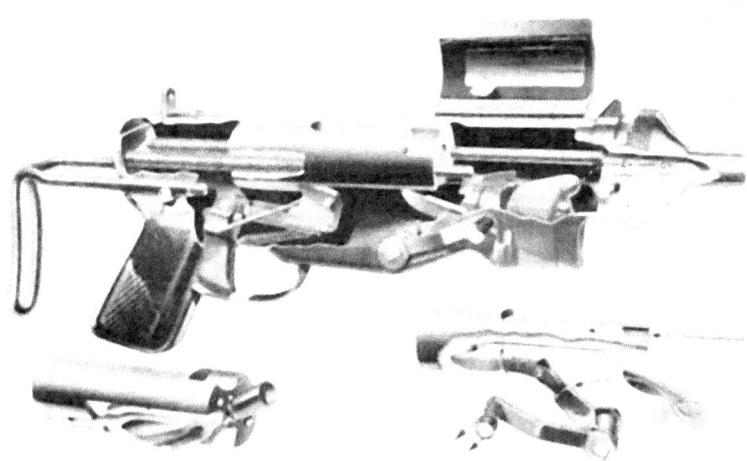

Figure 32. M3, cutaway view. Lower left shows camming action of ejector on empty cartridge case. Lower right shows action of retracting lever and handle in cocking the gun.

c. If the trigger is released, the nose of the sear engages in the sear notch in the bottom of the bolt, and holds the bolt in its rearward or cocked position. The trigger spring furnishes the spring action to the sear. The front arm of the trigger pin is a stop for the sear and prevents it from rotating forward when it engages the bolt.

19. Feeding

a. When a loaded magazine is placed in the weapon, the magazine catch holds the magazine in position. The top cartridge is held against the lips of the magazine through the action of the magazine spring and follower. When the bolt moves forward, it removes the round from the magazine.

b. When the bolt moves to the rear and clears the top of the magazine, the next cartridge is placed against the lips of the magazine by the action of the magazine spring and follower.

20. Chambering

The bolt, moving forward under the action of the expanding driving springs, pushes the top cartridge out of the magazine. The lips of the magazine aid to align the cartridge with the chamber. As the bolt continues forward, the cartridge is pushed into the chamber by the front of the bolt and the extractor. The base of the cartridge protrudes slightly from the chamber when the cartridge is fully seated.

21. Firing

After the cartridge is chambered, the bolt continues to move forward. The extractor springs out to the side and snaps into the extracting groove of the cartridge. At the same time, the fixed firing pin in the center of the cartridge recess of the bolt strikes the primer of the cartridge, firing the cartridge. At the instant of firing, the cartridge is inclosed in the chamber by the cartridge recess of the bolt, and the rim of the cartridge is engaged by the extractor.

22. Extraction

a. When the cartridge is fired, the gas pressure forces the bullet out of the muzzle and the empty cartridge case out of the chamber, pushing the bolt to the rear. The extractor holds the base of the cartridge case against the bolt. The bolt continues moving to the rear, carrying the empty cartridge case with it. Extraction is completed when the front of the cartridge case clears the rear of the chamber.

b. If the cartridge is not fired, the extractor will remove it from the chamber when the bolt is manually pulled to the rear.

23. Ejection

As the bolt moves to the rear, the empty cartridge case is held by the extractor. The base of the cartridge strikes the fixed ejector. The extractor serves as a pivot point for the cartridge, which is deflected out of the ejection opening of the receiver. The extractor and ejector are both needed to complete the ejection.

24. Cocking

As the bolt moves to the rear, the driving springs are compressed. If the trigger has been released, the nose of the sear will move up. As the bolt moves forward, the sear nose will engage in the sear notch and hold the bolt to the rear in a cocked position. If the trigger has not been released, the bolt will continue forward and the cycle of operation will be repeated.

Caution. If the gun is accidentally dropped, the bolt may be jarred far enough to the rear to clear the top cartridge in the magazine, but not far enough for the sear nose to engage in the sear notch. When this happens, the bolt will chamber and fire the cartridge as it goes forward.

25. Operation of the Housing Assembly

a. On the M3, when the retracting handle is pulled to the rear, the retracting lever pawl rises into the pawl notch in the bottom of the bolt. As the retracting handle is moved farther to the rear, the retracting lever pawl pushes the bolt to the rear until the bolt is engaged by the sear and held in a cocked position.

b. On the M3A1, the firer retracts the bolt with his finger. The retracting mechanism has been eliminated.

26. Operation of Safety Lock

a. When the bolt is forward and the cover is closed, the safety lock on the cover engages in the safety lock recess in the bolt. This prevents movement of the bolt.

b. When the bolt is to the rear and the cover is closed, the safety lock enters the cocking slot of the M3A1 bolt, or the notch on the front top portion of the M3 bolt, forces the bolt back off of the sear, and holds it to the rear. Closing the cover is called "locking the piece."

Section III. OPERATION

27. General

Before firing the submachine gun, the firer must know how to fill the magazine; must know how to load, fire, and unload the weapon; and must observe safety precautions. These points are covered in this section.

28. To Fill Magazine

a. Place the magazine loader on top of the magazine, then place the base of the magazine on a firm surface. Push down on the loader to depress the magazine follower. Insert a cartridge, base first, into the magazine. Lift the loader, and push the cartridge all the way into the magazine. Push down on the loader, depressing the cartridge and magazine follower. Repeat the operation (fig. 33) until the magazine is full.

b. To use the stock as the hand loader (fig. 34), place the butt of the stock over the magazine and use the same procedure as with the magazine loader.

29. To Load Submachine Gun

To load the submachine gun, pull the bolt sharply to the rear (cock),

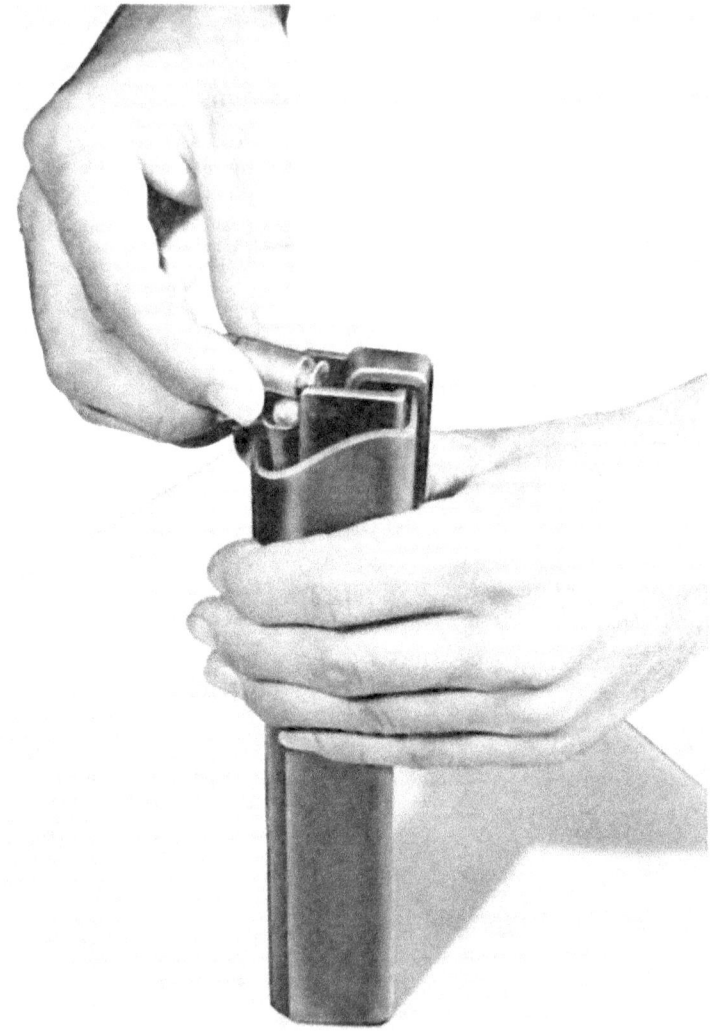

Figure 33. Using magazine loader.

close the cover (lock), insert the magazine, and push it upward until the magazine catch clicks into the magazine notch (load).

30. To Fire Submachine Gun

a. To fire the gun, raise the cover and manipulate the trigger.

b. The gun has no mechanism for semiautomatic fire. However, it is possible to fire single shots by proper manipulation of the trigger,

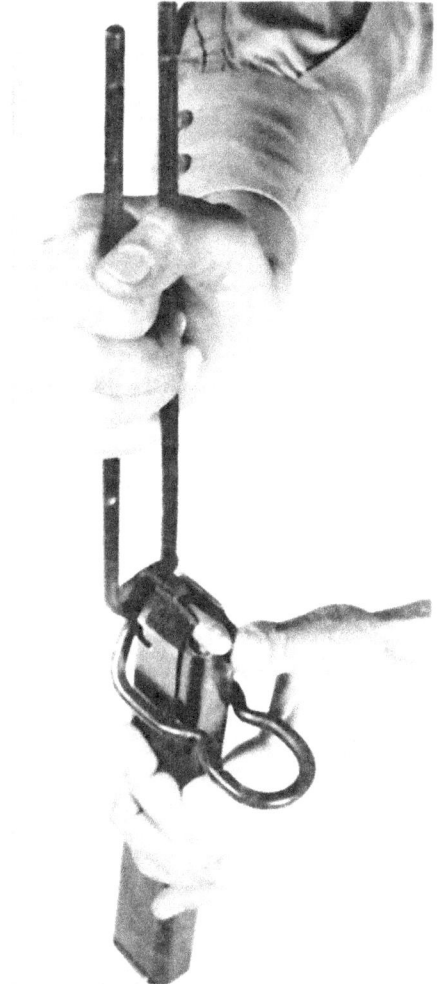

Figure 34. Use of the stock as a hand loader.

pressing it and then quickly releasing it. It takes practice for a firer to become proficient at firing single shots.

c. When firing long bursts, the weapon has a tendency to move to the right. The firer can control this tendency by always taking a correct firing position (par. 72).

d. When the magazine has been emptied, the bolt will close on the empty chamber. Cock and lock the weapon before inserting another loaded magazine.

31. To Unload Submachine Gun

Remove the magazine, and raise the cover. If the bolt is forward, pull it to the rear. Inspect the chamber (look and feel). Press the trigger and allow the bolt to go forward, then close the cover.

32. Safety Precautions, General

Safety cannot be overemphasized. The submachine gun has no mechanical means of locking the trigger. The insertion of a loaded magazine loads the gun. If the cover is open and the bolt cocked, pressure on the trigger will fire the gun. If an unlocked gun is dropped, it may fire whether the bolt is cocked or not. The safety precautions to observe in handling the submachine gun are:

  a. Never consider the weapon to be safe unless it has been properly cleared.

  b. Never playfully or carelessly point the weapon at anyone.

  c. Load the weapon only when ready to fire.

  d. Unlock the loaded weapon only when it is raised for firing.

  e. Never leave any obstruction in the muzzle or bore.

Section IV. MALFUNCTIONS, STOPPAGES, AND IMMEDIATE ACTION

33. General

A malfunction is a failure of the weapon to function satisfactorily. A stoppage is any unintentional interruption in the cycle of operation. If the submachine gun stops firing through no fault or intention of the firer, or an attempt to fire is made and the weapon does not fire, then a stoppage has occurred. The firer must be able to reduce a stoppage and continue firing. In combat, lives and the success of a mission may depend on the soldier's ability to reduce a stoppage quickly and continue to deliver accurate fire. Immediate action is the prompt action taken by the firer to reduce a stoppage.

34. Malfunctions

  a. Failure to Function Freely. Sluggish operation of the gun is usually due to excessive friction caused by dirt, lack of proper lubrication, burred or bent guide rods, or a dent in the receiver.

  b. Uncontrolled Automatic Fire (Runaway Gun). Uncontrolled automatic fire is fire that continues after the trigger has been released. This may be caused by the following:

  (1) A worn sear nose.
  (2) A worn sear notch on the bolt.
  (3) A weak or broken trigger spring.

In case of uncontrolled automatic fire, keep the gun pointed at the target and press the magazine catch to release the magazine.

35. Stoppages

a. Stoppages are classified in accordance with the six steps in the cycle of operation of the submachine gun (par. 16*c*). Stoppages are usually the result of worn parts or improper care of the gun. A knowledge of how the gun functions enables the soldier to classify and correct the stoppage. Listed below are the classes of stoppages which might occur.

    (1) *Failure to feed.* The top cartridge in the magazine is not positioned up and in front of the bolt. Most stoppages of the submachine gun are failures to feed caused by a defective or dirty magazine.

    (2) *Failure to chamber.* The top cartridge from the magazine is not seated in the chamber.

    (3) *Failure to fire.* The cartridge is chambered but does not fire.

    (4) *Failure to extract.* If the cartridge fires, the chamber pressure will usually push the empty cartridge case out of the chamber. If the cartridge case is not completely removed from the chamber and the bolt is retracted, then there is a failure to extract. This stoppage seldom occurs.

    (5) *Failure to eject.* The empty cartridge case is not ejected from the receiver.

    (6) *Failure to cock.* If the bolt is retracted and is not held by the sear, or if, during firing, the bolt does not move to the rear far enough to clear the top cartridge in the magazine, the gun has a failure to cock.

b. Common Stoppages. The two most common stoppages are:

    (1) Failure to feed—usually caused by a defective magazine.

    (2) Failure to fire—usually caused by defective ammunition.

c. Causes of Stoppages. The following chart lists common causes of various stoppages.

Stoppage	Cause	How to reduce
Failure to feed.	Dirty or dented magazine	Replace magazine.
	Weak or broken magazine spring	Replace magazine.
	Worn magazine notch	Replace magazine.
	Corroded ammunition	Replace ammunition.
	Worn or broken magazine catch	Replace magazine catch.
Failure to chamber.	Dirty chamber	Clean chamber.
	Obstruction in chamber	Remove obstruction.
	Weak driving springs	Replace driving spring.

Stoppage	Cause	How to reduce
Failure to fire.	Defective ammunition	Replace ammunition.
	Defective firing pin	Replace bolt.*
	Weak driving springs	Replace driving springs.
Failure to extract.	Broken extractor	Replace extractor.
Failure to eject.	Broken ejector	Replace ejector*.
	Broken or missing extractor	Replace extractor.
Failure to cock.	Worn sear	Replace sear*.
	Worn sear notch	Replace bolt*.
	Bent guide rods	Straighten.

* These items are not carried as unit repair parts.

d. Prevention of Stoppages. Periodic inspection and proper care and cleaning will reduce the possibility of the submachine gun having a stoppage.

36. Immediate Action

a. As the first step in reducing a stoppage, remove the magazine, retract the bolt, and inspect the chamber to insure that it does not contain a live cartridge or any other obstruction. If there is no obstruction, close the cover, replace the magazine, open the cover, and attempt to fire. If the gun still does not fire, check to see whether a live cartridge has chambered; if it has not, remove the magazine and insert a new magazine.

b. If there is a live cartridge or other obstruction lodged in the chamber, cock the gun and hold the cover down firmly; remove the barrel; then clear the chamber by using the stock to push the obstruction out of the barrel. Under combat conditions, when time is short, omit the step of removing the barrel.

Section V. CARE AND CLEANING

37. General

The submachine gun will function under conditions that would cause some automatic weapons to fail. However, its continued dependability and accuracy depend on its receiving proper care and cleaning. The chamber and bore, receiver, and moving parts must be kept clean and *very lightly* oiled. The same care must be given the magazines.

38. Cleaning Materials, Lubricants, and Rust Preventives

a. Cleaning Materials.

    (1) Cleaning compound, solvent (rifle-bore cleaner), is used to clean the bore and the face of the bolt after firing. It dissolves corrosive primer salts and removes primer fouling,

powder ash, and carbon. This cleaner has preservative properties and provides temporary protection against rust.

Caution. Rifle-bore cleaner is usable at temperatures of minus 20° Fahrenheit and higher. When it is below that temperature, it must be thawed and shaken well before it is used. *Do not mix water with rifle-bore cleaner.* This destroys its preservative qualities and impairs its value as a cleaner.

(2) Hot or cold water may be used to clean the bore when rifle-bore cleaner is not available. Hot, soapy water is preferable. One-quarter pound of soap dissolved in one gallon of water makes a desirable solution. After using the solution, dry the barrel thoroughly and apply a thin coat of oil.

(3) Volatile mineral spirits paint thinner and dry-cleaning solvent are noncorrosive solvents used for removing grease, oil, or light rust-preventive compounds from weapons. Apply them with rags to large parts, and use them as a bath for small parts.

Caution. These solvents are highly flammable. Do not smoke when using them. Continuous contact with them will dry the skin and may cause irritation.

(4) Decontaminating agents are used under special conditions to remove chemical agents (par. 45).

b. *Lubricants.*

(1) Medium preservative lubricating oil is a highly refined, non-hardening mineral oil containing a rust-inhibiting additive. It forms a relatively heavy film that resists direct action of salt spray. This makes it useful for coating all parts of a weapon before amphibious operations. It should be used in preference to special preservative lubricating oil only when the weapon is exposed to salt water, high humidity, or high temperatures. This oil should not be used in temperatures below freezing.

(2) Special preservative lubricating oil is a thin oil used for lubricating at normal and below-normal temperatures, and for providing temporary protection against rust. When this oil is used, moving parts of weapons must be checked frequently to make sure that they have an adequate film of lubricant.

(3) Engine oil SAE 10 may be used when preservative lubricating oils cannot be obtained. In cold weather, any heavy oil will cause sluggish operation, and may prevent the submachine gun from functioning properly. Engine oil does not have the rust-preventive properties of preservative lubricating oils. When engine oil is used, the weapon must be inspected, cleaned, and reoiled frequently.

c. Rust Preventive. Medium rust-preventive compound can be used to protect the submachine gun for a long period of time when it is boxed for storage.

39. Care and Cleaning Before Firing

Before the submachine gun is fired, the following steps should be taken to make sure that it will function properly:

a. Field disassemble the weapon.

b. Clean the bore and chamber with a clean, dry patch.

Note. Do not apply oil to the bore or chamber before firing.

c. Clean all parts thoroughly.

d. Use a lightly oiled cloth, apply a light coat of lubricating preservative oil to all parts that do not come in contact with the ammunition. Apply a light coat of rifle lubricant grease to the guide rods.

e. Assemble the weapon.

f. Wipe excess oil from the receiver.

g. Clean the magazines, and place a light film of oil on their outer surfaces.

40. Care and Cleaning During Firing

a. During lulls in firing, lubricate the guide rods, oil the sear pin, and trigger pin. Use the stylus of the oiler to apply oil from the oiler.

b. If time permits, during a lull in firing, disassemble the submachine gun and oil the sear, sear notch, connector pin, connector rivet, and the grooves in the bottom of the bolt.

41. Care and Cleaning After Firing

The weapon must be cleaned as soon as practicable on the day of firing and for the next three days, or longer if necessary, in the following manner:

a. Disassemble the groups.

b. Clean all parts with dry, clean cloth, using rifle-bore cleaner if necessary. Inspect all parts, and apply a light film of oil.

c. Clean the bore and chamber, using the following procedure:

(1) Saturate a patch with rifle-bore cleaner, and run it back and forth through the bore.

(2) Repeat the operation two or three times with clean patches saturated with bore cleaner.

(3) Run dry patches through the bore until they come out dry and clean.

(4) Inspect the bore for cleanliness. If it is not free of all residue, repeat the cleaning process. If the residue cannot be removed by the use of patches, the bore brush should be used.

(5) Clean the chamber with rifle-bore cleaner applied to a patch on the chamber cleaning brush.
(6) Dry the chamber, and inspect it for cleanliness.
(7) When the chamber and bore are both thoroughly clean, coat them with rifle-bore cleaner and leave them overnight.
(8) Clean the bore and chamber each day for the next three days, using the above procedure.
(9) After the bore and chamber are cleaned on the third day, if there is no sign of sweating, dry them and apply a light coat of oil.
(10) If there are signs of sweating, repeat the cleaning procedure each day until these signs disappear. Then dry the bore and chamber and apply a light coat of oil.

d. Assemble the weapon.

e. Perform an operation check (par. 15).

42. Care and Cleaning During Combat

a. There is no basic difference in the care and cleaning of the submachine gun during range firing and during combat. However, when the weapon is being fired each day, rifle-bore cleaner may be left in the bore and chamber each night.

b. To obtain the maximum efficiency from the submachine gun:
(1) Before firing, carefully check the bore and chamber for obstructions.
(2) Keep the bore and chamber free from oil and dirt when firing.
(3) Never leave a patch, plug, or other obstruction in the chamber or bore. Neglect of this precaution may result in serious injury.
(4) Oil the guide rods frequently to insure smooth operation of the gun.
(5) In emergencies, when the prescribed lubricants are not available, use any clean light mineral oil such as engine oil.

43. Preparation for Storage

Medium preservative lubricating oil is the most suitable oil for short-term protection. It is effective for storage over periods of two to 6 weeks, depending on climatic conditions. However, submachine guns in short-term storage must be inspected every 4 or 5 days and reoiled if necessary. For longer periods of storage, submachine guns are protected with medium rust-preventive compound. This compound is a semisolid material. It will preserve the weapon for a period of approximately one year, depending on climatic and storage conditions.

a. The weapon must be cleaned and prepared for storage with great care. The bore, receiver, and other parts of the gun should be thoroughly cleaned and completely dried with rags. In damp climates, particular care must be taken to see that the rags are dry. After a metal part is dried, it should not be touched with bare hands. All metal parts should then be coated with either medium preservative lubricating oil or medium rust-preventive compound, depending on the length of storage required. The best way to apply rust-preventive compound to the bore and chamber is to dip a clean bore brush into the compound and then run it through the bore two or three times. When the weapon is placed in storage, the bolt should be in the forward position.

b. Guns should be stored in wooden packing boxes which are provided with wooden supports for the stocks and muzzles of the guns. These supports should be coated with medium rust-preventive compound. Guns should be handled with oily rags while being placed in storage. Under no circumstances should a submachine gun be placed in storage in a cloth or other cover, or with a plug in the bore; this will cause the weapon to rust. The sling should be removed from the weapon and wrapped in oil paper coated with medium rust-preventive compound.

44. Cleaning Weapons Received from Storage

Submachine guns removed from storage will be coated with either preservative lubricating oil or rust-preventive compound. Weapons received from ordnance storage will usually be coated with rust-preventive compound. Use dry-cleaning solvent or volatile mineral spirits paint thinner to remove the compound or oil. Failure to thoroughly clean the driving spring recesses in the bolt may cause a malfunction or stoppage at below-normal temperatures, since the rust-preventive compound will congeal during cold weather. After using solvent, dry all parts by wiping them with a dry cloth. Then apply a thin film of special preservative lubricating oil.

45. Care and Cleaning in Connection with a CBR Attack

a. General. If a chemical, biological, or radiological attack is anticipated, the following action is taken: Apply oil to all outer metal surfaces of the submachine gun and accessories. *Do not* apply oil to ammunition. If the weapon is not to be used, cover the weapon, accessories, and ammunition with protective coverings and place them under natural cover. Ammunition should be kept in its containers as long as possible.

b. After a CBR attack, determine by means of detectors whether or not the equipment is contaminated. A complete suit of protective clothing, including protective gloves and a gas mask, must be worn during decontamination. If the contamination is too great, it may be necessary to discard the equipment. Detailed information on decontamination is contained in FM 21-40 and TM 3-220.

46. Care and Cleaning Under Unusual Climatic Conditions

a. Cold Climates. It is necessary that the moving parts of the weapon be kept absolutely free from moisture. Excessive oil on the working parts will solidify and cause sluggish operation or complete failure.

(1) Before firing in temperatures below 0° Fahrenheit, disassemble the gun and clean all parts of the gun and magazine thoroughly. Oil them very lightly by rubbing them with a cloth dipped in special preservative lubricating oil. Leave the bore and chamber free of oil.

(2) When the gun is brought indoors, allow it to come to room temperature; then disassemble it, wipe it completely dry of any moisture, clean it, and oil it lightly with special preservative lubricating oil.

(3) If the gun has been fired, the bore should be immediately swabbed out with an oily patch. When the weapon reaches room temperature, clean and oil it as prescribed in paragraph 41.

b. Tropical Climates. Where temperature and humidity are high, or during rainy seasons, thoroughly inspect the weapon daily and keep it lightly oiled when not in use. Remove the groups at regular intervals and, if necessary, disassemble them for cleaning, drying, and oiling. Be careful to see that all unexposed parts, as well as exposed surfaces, are kept clean and oiled with special or medium preservative lubricating oil.

c. Hot, Dry Climates.

(1) In hot, dry climates, where sand and dust are likely to get into the mechanisms and bore, the weapon should be wiped clean daily, or oftener if necessary. Groups should be disassembled to insure thorough cleaning.

(2) When the weapon is used under sandy conditions, lubricants should be wiped from exposed and noncritical operating surfaces. This will prevent sand or dust from sticking to the lubricants and forming an abrasive which can damage the moving parts.

(3) Immediately after use in sandy terrain, the weapon should be cleaned and lubricated with special preservative lubricating oil.

(4) During sand or dust storms, the gun should be kept covered. It should be cleaned immediately after the storm.

Section VI. REPAIR PARTS AND ACCESSORIES

47. Repair Parts

a. The parts of any submachine gun may in time become unserviceable through breakage or wear resulting from usage. For this reason, extra parts are provided for replacement of parts most likely to fail. Sets of repair parts must be kept complete at all times; when a part is used, it should be replaced in the set as soon as possible. Repair parts are kept clean and lightly oiled to prevent rust. Parts must always be ready for immediate use.

b. Extra 30-round magazines are also issued with the gun. The number of magazines issued per gun, and the allowance of repair parts, are prescribed in appropriate supply publications.

48. Accessories

Accessories include the tools required to disassemble and assemble the submachine gun, cleaning and preserving materials, sling, repair parts envelope, oiler, flashhider, magazine loader, and similar items. They must be used for no other purpose than that for which they are intended.

Section VII. AMMUNITION

49. General

The soldier must be able to recognize the types of ammunition used in the submachine gun in order to take proper care of the ammunition and to select the proper type for a specific purpose.

a. A live cartridge for a submachine gun consists of a cartridge case, primer, propelling charge, and bullet.

b. The term *ball ammunition* refers to a cartridge having a bullet that consists of a metallic jacket filled with lead.

50. Classification of Ammunition

Based upon use, the principal classifications of ammunition for the submachine gun are—

a. Ball—for use against personnel and for guard duty.

b. Tracer—for observation of fire and for incendiary and signal purposes.

c. Dummy—for training.

51. Ammunition Lot Number

When ammunition is manufactured, it is given an ammunition lot number. This lot number is marked on all packing containers. It appears on all records pertaining to the ammunition. It must be given in all reports on the condition and functioning of the ammunition, or on accidents in which the ammunition is involved. Therefore, it is important to retain the lot number with the cartridges after they are removed from their original containers. If cartridges cannot be identified by ammunition lot number, they are automatically placed in grade 3. Grade 3 ammunition is unserviceable; it will not be fired, but will be turned in to the issuing ordnance officer.

52. Identification of Ammunition

a. Markings. The contents of original boxes or containers can be readily identified by markings on the box. These markings indicate the number of cartridges in the box or container, the caliber, the type, the code symbol, and the lot number.

b. Identification of Ammunition Types. When removed from their boxes, cartridges can be identified, except for the ammunition lot number and grade, by physical characteristics as described below.

   (1) *Ball.* The bullet of the ball cartridge has a jacket of gilding metal, gilding-metal-clad steel, or copper-plated steel.
   (2) *Tracer.* The bullet of the tracer cartridge has a copper-plated steel or gilding-metal-clad steel jacket, which is painted red for a distance of approximately 3/16 of an inch from the tip.
   (3) *Dummy.* This cartridge was formerly identified by its tinned case, which either had an inert primer with holes drilled in the cartridge case or had no primer. Since 1944, dummy cartridges have been manufactured without a tinned case; they can be identified by the empty primer pocket and two holes drilled in the side of the case. The bullet in both cartridges is the same as that used in the ball cartridge.

53. Care, Handling, and Preservation of Ammunition

a. Small arms ammunition is generally safe to handle. However, do not allow ammunition boxes to become broken or damaged. Repair broken boxes immediately. Transfer all original markings to the new parts of the box.

b. Do not open ammunition boxes until the ammunition is to be used. Ammunition removed from airtight containers, particularly in damp climates, is likely to corrode, thereby becoming unserviceable.

c. Be careful when opening wooden ammunition boxes, because they are used as long as they are serviceable.

d. Protect ammunition from mud, sand, dirt, and water. If it gets wet or dirty, wipe it off at once, with a clean dry cloth. Wipe off light corrosion as soon as it is discovered. Cartridges with a heavy coat of corrosion must be turned in to the issuing ordnance officer.

e. Do not oil or polish cartridges.

f. Do not expose ammunition to the direct rays of the sun for any length of time. If the powder is heated, excessive pressure will be developed when the weapon is fired. This condition will affect accuracy and the operation of the weapon.

g. Do not attempt to fire cartridges that have dents, scratches, loose bullets, or rusted cases. If a cartridge is defective, turn it in. Do not throw away or attempt to destroy defective ammunition.

h. Do not strike the primer of a cartridge; it may ignite and cause injury.

54. Storage of Ammunition

a. Small arms ammunition is not an explosive hazard. Under poor storage conditions it may become a fire hazard.

b. Small arms ammunition should be stored away from all sources of extreme heat.

c. Whenever practicable, small arms ammunition should be stored under cover. If it is necessary to leave ammunition in the open, it should be raised on dunnage at least six inches from the ground. It should be covered with a double thickness of tarpaulin or suitable canvas. The cover should be placed so that it gives maximum protection yet allows free circulation of air. Suitable trenches must be dug to prevent water from flowing under the ammunition pile.

55. Precautions in Firing Ammunition

The general precautions concerning the firing and handling of ammunition in the field, as prescribed in AR 385-63 and TM 9-1990, must be observed. Precautions particularly applicable to small arms ammunition include the following:

a. No small arms ammunition will be fired until it has been positively identified by ammunition lot number and grade.

b. Small arms ammunition graded and marked "For Training Use Only" will not be fired over the heads of troops under any circumstances.

c. Before firing, the firer must be sure that the bore of the weapon is free from any foreign matter such as cleaning patches, mud, sand, or snow. Firing a weapon with any obstruction in the bore will result in damage to the weapon and possible injury to the firer.

d. When a bullet lodges in the bore, it should be removed by the application of pressure from the muzzle end (par. 36). Attempting to shoot the bullet out with another cartridge is prohibited.

56. Hangfires

A hangfire is a temporary failure or delay in the action of a primer or propelling charge. When a hangfire occurs in any lot of ammunition, the use of the entire lot should be suspended and a report made to the post ordnance officer, giving the lot number involved. The lot affected will be withdrawn and replaced with serviceable ammunition.

CHAPTER 3

MANUAL OF ARMS

57. General

The manual of arms for the submachine gun is designed to provide uniform, simple, safe, and quick methods for handling the gun. This chapter gives a simple and effective method of performing the manual of arms, which can be executed in cadence when precision is desired.

58. Carrying Position

The submachine gun is carried with the magazine removed and stock telescoped, slung over the right shoulder, muzzle pointing down, with the right hand grasping the sling in front of the armpit (fig. 35). When carried during dismounted marches or during field exercises, the weapon may be slung over either shoulder. When troops are at ease, the submachine gun is kept slung, unless otherwise ordered. When troops are at rest, the submachine gun may be unslung and held in any desired position. In executing the command ATTENTION, the soldier assumes the position of attention and places the submachine gun in the carrying position. Parade rest is executed in the normal manner except that the right hand continues to grip the sling and keeps the gun at sling arms.

59. Port Arms

a. The command is PORT, ARMS.

b. Using the sling, swing the gun forward with the left hand and grasp the stock with the right hand. Withdraw the left hand from the sling, and grasp the housing assembly with the left hand. Extend the stock, then carry the gun to a position four inches in front of the body, with the barrel pointing upward and to the left at an angle of 45° and with the barrel collar at the same level as the point of the left shoulder (fig. 35). Hold the gun in a vertical plane parallel to the body, with the left hand at the housing assembly and the right hand grasping the butt of the stock.

60. Inspection Arms

a. The command is INSPECTION, ARMS.

b. First execute port arms. With the right hand, open the cover, retract the bolt, and grasp the butt of the stock with the right hand.

At the same time, lower the head and eyes enough to look into the receiver. (Be sure that the fingers of the left hand do not cover the ejection opening (fig. 35)). Having seen that there is no round in the chamber, raise the head and eyes to the front. At the command PORT, ARMS, remove the right hand from the stock, press the trigger, close the cover, and grasp the butt of the stock with the right hand.

61. Present Arms

a. The command is PRESENT, ARMS.

b. Grasp the sling in front of the armpit with the left hand, and give the hand salute with the right hand (fig. 35). At the command ORDER, ARMS, drop the right hand smartly to the side, raise the right hand and grasp the sling in front of the armpit, then drop the left hand to the side.

62. Sling Arms

a. Sling arms is executed on the command ORDER (or RIGHT SHOULDER), ARMS after INSPECTION, ARMS and PORT, ARMS have been given.

b. At the command of execution, telescope the stock, grasp the receiver with the left hand, thrust the right arm through the sling, and assume the carrying position.

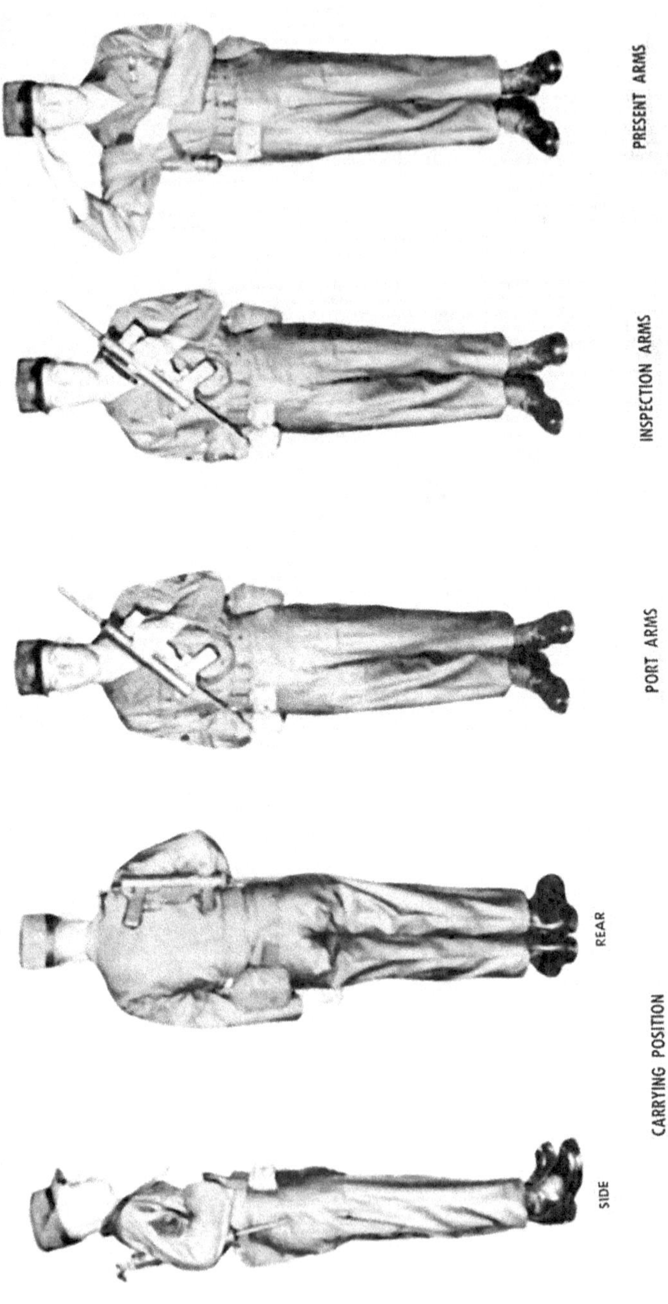

Figure 85. Manual of arms.

CHAPTER 4

MARKSMANSHIP TRAINING

Section I. GENERAL

63. Introduction

The primary use of the submachine gun is to engage the enemy at close range with accurate automatic fire. To obtain accurate fire from the submachine gun, the soldier must be trained. With proper training and a desire and willingness to learn, the soldier can deliver very effective fire with this weapon.

64. Fundamentals of Marksmanship

Accurate shooting is the result of knowing and being able to put into use the important elements of marksmanship: sighting and aiming, positions, and trigger manipulation.

65. Phases of Training

a. Marksmanship training is divided into two phases—
  (1) Preparatory marksmanship training.
  (2) Range firing.

b. Each of the two phases may be divided into separate instructional steps. One very important thing to remember, during all phases of marksmanship training, is that the training must be progressive.

Section II. PREPARATORY MARKSMANSHIP TRAINING

66. General

a. Before he receives instruction in marksmanship training, the soldier must have a good understanding of functioning. He must know how to disassemble and assemble the weapon. He must know the correct way of applying immediate action and all of the safety precautions.

b. A thorough course in preparatory marksmanship training must precede any range firing. This training is given to all soldiers expected to fire the submachine gun during range practice, including those who have previously qualified. The soldier should develop fixed and correct shooting habits before going on the range. The purpose of preparatory marksmanship training is to develop these shooting habits.

c. Preparatory marksmanship training is divided into five phases, which should be taught in the following order:

(1) Sighting and aiming exercises.
(2) Position exercises.
(3) Trigger manipulation exercises.
(4) Marksmanship exercises.
(5) Examination.

67. Coaching

a. Throughout all of the preparatory training, the coach-and-pupil method of training should be used. The duties of the coach are very important. How well man learns to shoot depends to a great extent on how well his coach does his duties. When possible, the more experienced man should serve as the coach first. The coach will assist the pupil by—

(1) Correcting any errors made.
(2) Insuring that he takes the proper positions.
(3) Insuring that he observes all safety precautions.

b. During range practice, the coach will perform the duties listed in paragraph 82.

68. Record of Instruction and Qualification

A record of the progress of each individual's training must be kept. This record will show what extra training he needs. It will also serve as a guide in the selection of new instructors or assistants. A sample form for keeping a record of individual progress in training is shown in figure 36. This form when reproduced comes under the provisions of paragraph 20, AR 310–1.

69. Sighting and Aiming, General

Sights on the submachine gun are not adjustable. The weapon is primarily intended for firing automatic fire at short ranges where quick shooting is required. Generally the sights are used for the initial alignment of the weapon on the target. During firing, the firer can observe the strike of the rounds and bring them onto the target. In a situation where the firer desires to fire single shots or short bursts and has enough time to obtain the correct sight alignment and sight picture, he should do so. Therefore he must understand the correct sight alignment and sight picture.

70. Sighting and Aiming Exercises

a. First Exercise. This exercise teaches the correct alignment of the front and rear sights, and the correct sight picture, when aiming at a target.

RECORD OF SUBMACHINE GUN INSTRUCTION

_____
(Army Serial Number)

_____ _____ _____
     (Last Name)   (First Name)   (Middle Initial)

Grade_____ Organization_____

PREPARATORY TRAINING AND RANGE FIRING

	DATE	RESULT	INITIALS	REMARKS
General description of weapon				
Disassembly				
Assembly				
Care and cleaning				
Functioning				
Stoppages and immediate action				
Sighting and aiming				
Positions: standing, sitting, kneeling, prone, assault				
Trigger manipulation				
Safety precautions				
Loading and unloading				
Ability to coach				
Examination				

METHOD OF MARKING

UNSATISFACTORY: X

SATISFACTORY: XX

HAS INSTRUCTIONAL ABILITY: XXX

① Front

Figure 36. Sample form for recording progress of training.

RECORD OF SUBMACHINE GUN INSTRUCTION
MARKSMANSHIP TRAINING AND RECORD FIRING

	SCORE	DATE	INITIALS
Familiarization firing			

MARKSMANSHIP EXERCISES

First _____

Second _____

Third _____

INSTRUCTIONAL PRACTICE	TARGETS HIT	NR OF HITS	SCORE	DATE	INITIALS
Dismounted practice					
Dismounted practice					
Dismounted practice					
Vehicular firing					
Vehicular firing					
Moving ground targets					

RECORD FIRING					SIGNATURE
Dismounted practice					

Qualification (dismounted practice course)

    Expert

    Sharpshooter

    Marksman

    _____
    (Signature of Organization Commander)

    _____
    (Grade and Organization)

② Back

Figure 36—Continued.

(1) The sighting and aiming bar (fig. 37) is used for this exercise. The front and rear sights of the sighting and aiming bar represent the sights of the submachine gun. The eyepiece represents the position of the firer's eye and is so constructed that everyone who looks through the eyepiece will see the same sight picture.

(2) The instructor or coach adjusts the sights to show the correct alignment, and then has each man look through the eyepiece. The sights are correctly aligned when the top of the front sight is seen in the middle of the rear sight.

(3) The coach then moves the sights out of alignment, and the pupil realigns them. The coach should demand precision in this realignment.

(4) When the pupil is thoroughly familiar with the alignment of the sights, the coach places the bull's-eye on the bar to complete the sight picture. When the sight picture is correct, the top of the front sight just touches the bottom of the bull's-eye. Each man looks at the sight picture through the eyepiece.

(5) The coach then moves the bull's-eye, and the pupil readjusts it to set up the correct sight picture.

(6) The coach sets up errors in sight alignment and sight picture and has the pupil detect the errors and correct them.

b. Second Exercise. In this exercise, the submachine gun is used to teach the correct sight alignment and sight picture.

(1) The submachine gun is placed in a rest and aimed at a blank piece of paper mounted on a box 50 feet away. One man sits on the box and acts as a marker. He is given a small black disk, mounted on a handle, to use as a bull's-eye. A small hole is punched in the center of the disk.

(2) The coach, without touching the gun or rest, takes a prone position and, looking through the sights, gives commands to the marker to move the disk until it is in correct alignment with the sights. He then calls HOLD to the marker. Each pupil looks through the sights to see the correct sight picture.

(3) The marker moves the disk. A pupil takes a prone position and directs the marker to move the disk until he has the correct sight picture.

(4) The coach sets up slight errors in the sight picture and requires each pupil to correct them.

c. Third Exercise. This exercise is designed to show the importance of uniform sight pictures.

(1) The exercise is the same as the second exercise, except that

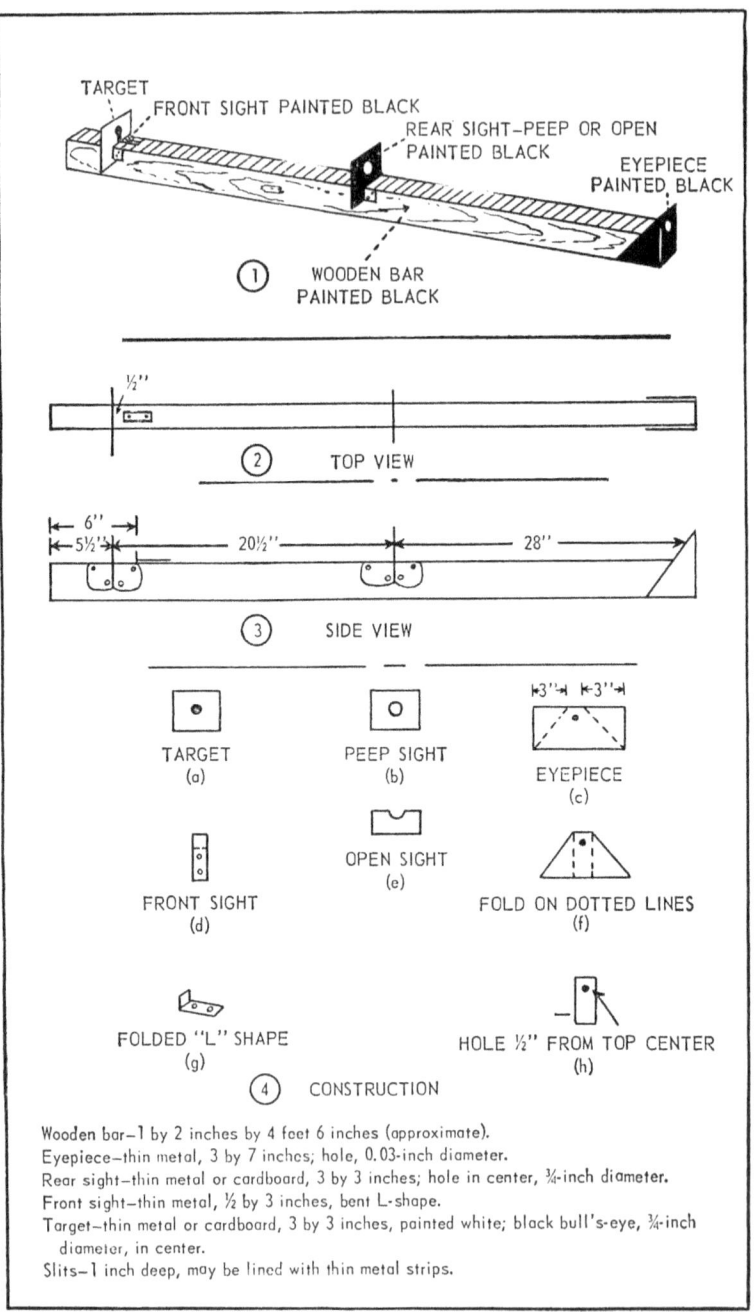

Figure 37. Sighting bar.

each pupil makes three sight pictures, and these pictures are plotted.
(2) The pupil directs the marker to move the disk until a correct sight picture is obtained. He then calls HOLD to the marker.
(3) The coach commands MARK. The marker places a dot on the paper by inserting the point of a pencil through the hole in the center of the disk. He numbers the dot.
(4) The marker moves the disk, and the procedure is repeated twice more.
(5) The coach comments on the shot group and explains how any errors may be corrected.
(6) Each pupil will repeat the exercise until he can cover his group with the eraser end of a pencil.

71. Important Points About Sighting and Aiming

a. The front sight must be accurately centered in the rear sight.

b. The bull's-eye is centered above and appears to barely touch the top of the front sight.

c. The last focus of the eye is on the *front sight*. The front sight will be seen clear and sharp, while the bull's-eye will appear to be a bit fuzzy.

d. If the rounds are striking below the target, aim higher on the target. If the rounds are striking above the target, aim lower.

72. Position

The second and most important phase of preparatory marksmanship is the position exercise. To hit a target and to continue to hold a burst on a target, the firer *must* have a good position. The submachine gun may be fired from the standing, sitting, kneeling, prone, or assault position (fig. 38).

a. Standing. This is normal firing position. To assume this position, stand facing the target, then make a half right face. Move the left foot forward one step, pointing the left toe toward the target. Lean forward; bend the left knee slightly, keeping the right leg straight, with about two-thirds of the body weight on the left foot. Grasp the magazine guide with the left hand and the pistol grip with the right hand. Place the butt of the stock against the right shoulder, and twist the body (at the waist) to the left to bring the right shoulder forward. The left elbow should be under the weapon, and the right elbow should be shoulder high. Press the cheek against the stock. The recoil is slight for one shot; but in automatic fire, each time the gun recoils it will tend to push the shoulder backwards. Therefore, the gun will move off the target if the firer is not well braced and in the proper position.

b. Sitting. This position is best used when firing from ground that slopes to the front. To assume this position, face the target, half face to the right, spread the feet a comfortable distance apart, and sit down. The feet should be farther apart than the knees. Bend the body forward from the hips, keeping the back straight. Push the right shoulder slightly forward (toward the target). Place the left upper arm on the flat part of the shinbone so that the tip of the elbow is crossed over the shinbone. There should be several inches of contact between the upper arm and the shinbone. The right elbow is blocked in front of the right knee.

c. Kneeling. The kneeling position affords a steadier aim than the standing position and is useful when the firer can crouch behind a rock, log, or other protection. This position is frequently used on level ground or ground that slopes upward. To assume this position, face the target, half face to the right, and kneel on the right knee. Sit on the right heel, with the right thigh forming an angle of 90° with the line of aim. The entire surface of the lower right leg, from knee to toe, is in contact with the ground. The left foot should be placed about 18 inches to the front, with the toe pointing at the target. The left lower leg is vertical when viewed from the front. Move the weight of the body forward, and place the point of the left elbow a few inches forward of the knee. The right elbow is raised to the height of, or slightly below, the right shoulder.

d. Prone. This position is the steadiest and should be used whenever time and terrain permit. To assume this position, take a prone position, with the body inclined to the left of the line of aim at an angle of 20° or less. Spread the legs a comfortable distance apart, with the toes pointed outward. Keep the spine straight. Place the left elbow under the gun, with the left hand grasping the magazine guide. The right elbow is out from the body so that the shoulders are level. Place the butt of the stock in the pocket formed by the shoulder, and press the cheek against the stock.

e. Assault. This position, usually called the hip position or chest position, is used for "close-in" fighting. When this position is used, there is less tendency of the muzzle to climb. The sights are not used to aim the weapon; the firer simply points the weapon toward the target and commences firing. The soldier must have a great deal of practice before he can do accurate shooting. To assume this position, press the stock against the side of the hip with the right arm, or place the stock under the armpit and press it against the body. The body should be in a crouched position, and the firer should walk on the balls of his feet so that he can quickly shift his body to fire at targets to his side.

73. Trigger Manipulation

a. With a semiautomatic weapon, the most important single factor

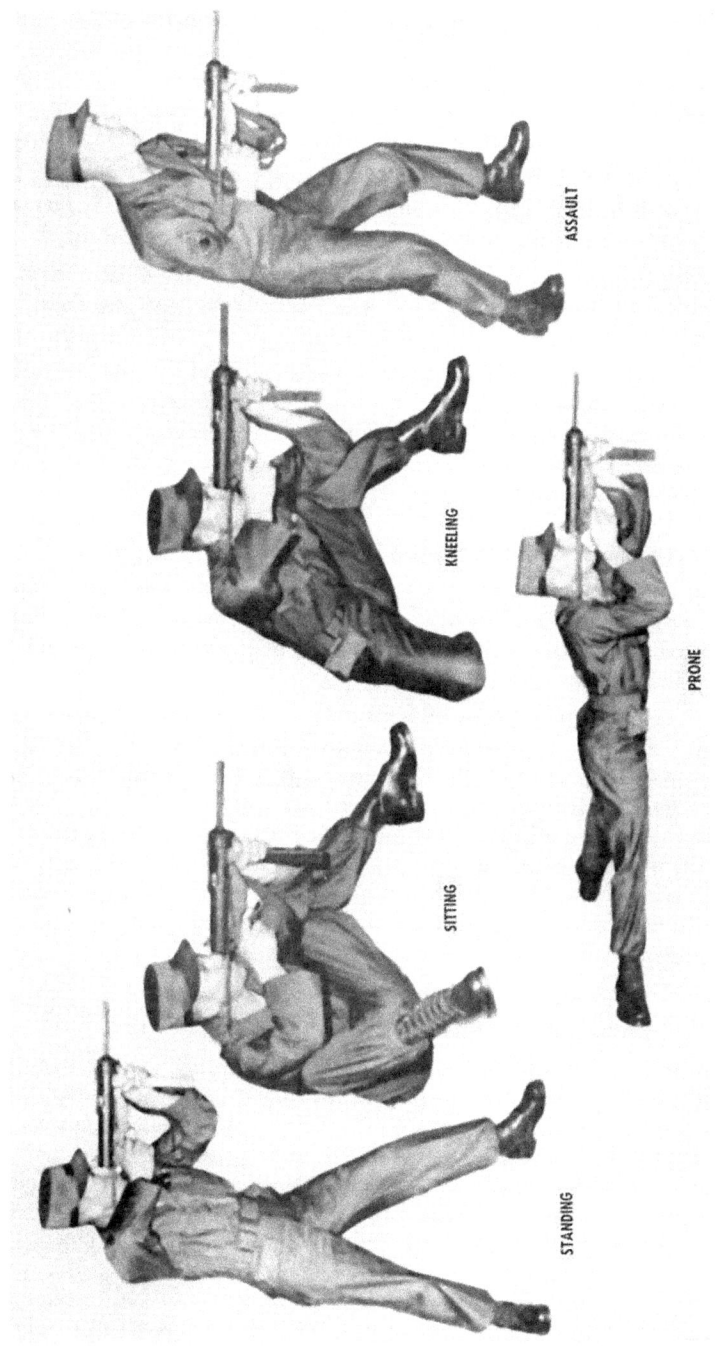

Figure 38. Firing positions.

in marksmanship is trigger squeeze. With an automatic weapon, the soldier does not squeeze the trigger; he manipulates the trigger to obtain a single shot or a burst of rounds. To obtain correct trigger manipulation on the submachine gun, the soldier must be given adequate training. In addition, he must be familiar with his weapon, because the trigger pull on all submachine guns is not the same.

b. To fire single shots, the firer may press or tap the trigger to the rear until the bolt is released, then quickly release the trigger.

c. To obtain a burst of rounds, the firer holds the trigger to the rear until the desired number of rounds have been fired, then quickly releases the trigger.

d. To practice trigger manipulation for single shot, the soldier cocks the weapon, presses or taps the trigger, and releases it before he hears the noise of the bolt striking the guide rod locating plate. If ammunition is available, firing the weapon and becoming familiar with it is the best means of learning trigger manipulation.

74. Nonfiring Marksmanship Exercises

The following simulated fire exercise teaches the soldier to spot his targets by any slight movement in his general field of vision. From the standing or assault position, he moves his gun rapidly to point at the targets as they appear.

a. The exercise is conducted on the dismounted practice course (par. 82 and fig. 39). No firing is simulated for phase A. The firer and coach stand at point Q; all targets are concealed from the firer. The firer holds the gun, which is cocked and unloaded. The firer is instructed to commence walking along the firing line and to look over his field of fire (phase B) by shifting his eyes, without focusing them on any particular object, until he sees a group of targets appear. The firer quickly points his gun at the group of targets and simulates firing from one of the following positions:

(1) Standing position, simulating aimed fire.

(2) Assault position, simulating unaimed fire.

From either position, he simulates firing a short burst at each target, shifting the weapon after simulating each burst. Without moving the gun from the position taken up, the firer continues walking along the course, looking over his field of fire until he sees another group of targets. The firer must stop walking to simulate firing. The targets are exposed by the target operator, who gives the firer no advance indication as to which group of targets will be exposed. This same procedure is carried out for phases C and D. Initially, targets are exposed for ten seconds. Each time the firer performs this exercise, the time that the targets remain exposed is reduced until the time is two seconds, except that the time for targets in groups 4 and 9 is reduced only to five seconds.

b. Other field targets or combat ranges (TM 9–855) may be used for marksmanship exercises. If the time and space are available, a course may be built where targets drop from trees, spring up out of foxholes, spring from behind trees, and spring up behind and to the sides of the firer. The more targets used and the more surprise, the better the training for the firer.

c. After the firer has fired the dismounted courses, he is trained in locating and firing at targets from a vehicle. He must practice the vehicular course (fig. 46) several times with simulated fire before he is permitted to fire the course.

75. Examination Before Range Firing

a. An examination is the last step of preparatory marksmanship training. It should be given to and passed by all men before they fire the submachine gun on the range.

b. The examination should be given far enough in advance of range firing to permit additional instruction to those men who made an unsatisfactory showing in any phase of training.

c. The record of submachine gun instruction (fig. 36) will serve as a progress chart during training.

d. During the oral or practical examination, the examiners take the names of the men who need extra instruction. These men are then given extra instruction in the subjects needed.

76. Examples of Questions for Examination

The questions listed below are given only as a guide. Any pertinent questions on the subjects listed on the record of submachine gun instruction (fig. 36) may be asked. Pupils may give their answers in writing, orally, or by demonstration.

Q. Name the parts of the weapon as I point to them.
A. Pupil names each part pointed out.

Q. How many rounds of ammunition can be placed in a magazine?
A. Thirty rounds.

Q. Show me the extractor, ejector, trigger pin.
A. Pupil points to each part as it is named.

Q. What is the maximum effective range of the weapon?
A. 100 yards.

Q. Demonstrate field disassembly and assembly.
A. Pupil dissassembles and assembles the weapon.

Q. How do you remove powder fouling from the bore?
A. By swabbing it thoroughly with patches saturated with rifle-bore cleaner or hot soapy water.
Q. In freezing weather, what type of oil should be used on the submachine gun?
A. Oil, lubricating preservative, special.
Q. What is the most important phase of preparatory marksmanship training?
A. Learning the correct positions.
Q. Demonstrate the standing, sitting, and assault positions.
A. Pupil demonstrates the required positions.
Q. You are on the range. What do you do at the command COCK, LOCK, AND LOAD?
A. Unlock the gun, cock it, lock the gun, and insert the magazine.
Q. Demonstrate how to use the hand loader.
A. Pupil demonstrates the use of the stock in loading a magazine.
Q. In case of a stoppage, what should the firer do?
A. Apply immediate action.
Q. Demonstrate how to apply immediate action.
A. Pupil demonstrates immediate action.
Q. Name the steps in the cycle of operation.
A. Feeding, chambering, firing, extraction, ejection, and cocking.
Q. While firing on the range, you have a stoppage. You notice that there is a spent cartridge case jammed between the bolt and the rear of the barrel. Classify the stoppage.
A. A failure to eject.

Section III. COURSES FIRED

77. General

a. All Army personnel are authorized to fire a qualification course with their basic weapon and a familiarization course with other weapons. See AR 370-5.

b. Range firing is started as soon as the soldier completes preparatory marksmanship training. Range firing consists of—

   (1) Instruction firing, which is practice firing on a range with the help of a coach.
   (2) Record firing, which is the final test of the soldier's proficiency and is the basis for his classification in marksmanship.
   (3) Familiarization firing, which acquaints the soldier with the weapon. It is fired only by soldiers who are not authorized to fire the record course for qualification.

c. The amount of instruction firing is not limited to that prescribed. Additional practice may be given as time and ammunition allowances permit.

d. Scores of all firing done with the submachine gun should be entered on the record of submachine gun instruction (fig. 36). Figure 47 shows a sample score card for instruction practice and record firing. Scores of familiarization and other types of firing may be entered directly on the record of submachine gun instruction, or score cards may be prepared by extracting the INSTRUCTIONAL PRACTICE portion of the record of submachine gun instruction.

78. Instruction Practice Firing

a. The following table prescribes the firing for the dismounted practice course (fig. 39). The table is fired three times for instruction and once for record. This course is designed to emphasize both alertness and accuracy. The procedure for firing the course is given in paragraph 82.

SLOW FIRE—TARGET M

Phase	Type of fire	Position	Range	Time	Rounds
A	Single shot and automatic	Standing, kneeling and prone	60 yards	No limit	*15

* Five rounds will be fired from each position. The five rounds from the prone position will be automatic fire.

QUICK FIRE—TARGETS E AND F

Phase	Type of fire	Position	Range	Time	Rounds
B	Single shot or automatic	Standing or assault	25–35 yards	Each group of targets exposed 2 seconds	15 rounds, 3 per target
C	Single shot or automatic	Standing or assault	20–40 yards	Groups 5 and 6 exposed 2 seconds; group 4 exposed 5 seconds	15 rounds, 3 per target
D	Single shot or automatic	Standing or assault	15–30 yards	Groups 7 and 8 exposed 2 seconds; group 9 exposed 5 seconds	15 rounds, 3 per target

Note. Due to the short time that the targets are exposed, the firer will have to fire most of the rounds by automatic fire.

b. Phase A is fired from a stationary position. Targets in phases B, C, and D are exposed while the firer is walking; the firer must stop walking to take the targets under fire.

c. Target M in phase A is a stationary target. Groups 1, 2, 3, 5, 6, 7, and 8 are bobbing type targets, and groups 4 and 9 are moving type targets.

79. Record Firing

a. The table used for instruction practice (par. 78) is fired once for record. Procedures to be followed for the conduct of the range and scoring are outlined in section IV.

b. Coaching is not permitted during record firing.

c. Qualification scores are as follows:

Expert _____ 180 points
Sharpshooter _____ 160 points
Marksman _____ 140 points

d. Record of scores should be entered on each individual's score card. A sample score card is shown in figure 47.

80. Familiarization Firing

The following table prescribes the familiarization firing. Preliminary instruction is conducted prior to firing. Thirty rounds of ammunition are allowed per individual. The procedure for familiarization firing is given in paragraph 83.

FAMILIARIZATION FIRING—TARGET M

Type of fire	Position	Range (yards)	Time	Rounds
Single shot and automatic	Standing and assault	25	No limit	*10
Single shot and automatic	Standing, sitting, or kneeling	45	No limit	**10
Single shot and automatic	Standing, sitting, or kneeling	60	No limit	**10

* Four rounds from the standing position (single shot) and six rounds in a burst from the assault position.
** Four rounds single shot and six rounds in bursts of three rounds.

Section IV. RANGE FIRING

81. Responsibility

Organization commanders are responsible for the conduct of range practice of their organizations in accordance with the provisions of this manual and applicable Army Regulations. All range firing will be conducted under the direct supervision of a commissioned officer. No person will be permitted to fire the submachine gun on the range until he has completed the preliminary instruction and has successfully passed the examination prescribed in preparatory training.

82. Procedure for Conducting Dismounted Practice and Record Firing

The officer in charge of firing, and four scorers, take positions on the firing line. There is one scorer at the starting point for each phase (points P, Q, S, and U) as shown in figure 39. The first firer reports to the scorer at point P, with four magazines loaded with 15 rounds in each. The officer in charge commands: 15 ROUNDS BALL AMMUNITION, COCK, LOCK, AND LOAD. At this command the firer takes a crouched position, raises the cover, retracts the bolt, closes the cover, inserts a magazine, and calls READY.

  a. *Phase A.*

   (1) As soon as the firer calls READY, he takes up the standing position, after which the officer in charge commands: FIRE WHEN READY.

   (2) At this command, the firer fires five rounds (single shot) at the M target. He then assumes the kneeling position and fires five rounds (single shot) at the target. He then assumes the prone position and fires a burst of five rounds at the target. There is no time limit for phase A.

   (3) After the firer has completed phase A and has cleared his weapon (checked by the scorer), he places his weapon on the stand. The officer in charge gives the order to score. The firer and scorer move forward to score the target. The score card is then given to the scorer for phase B.

  b. *Phase B.*

   (1) When the first firer has completed phase A, he moves to a position on the firing line just short of the red flag designating the firing zone of phase B (point Q, fig. 39). The second firer moves up to the firing point for phase A with four magazines loaded with 15 rounds in each.

   (2) At the command of the officer in charge, both men take a crouched position; cock, lock, and load their guns; and announce READY. The officer in charge commands: FIRE WHEN READY.

   (3) At this command, the firer at phase A (second firer) takes a standing position and commences firing phase A. The firer at point Q (first firer) walks along the phase B firing line and fires at group 3 and group 1 or 2, as the targets appear. The target operator will expose either group 1 or group 2,

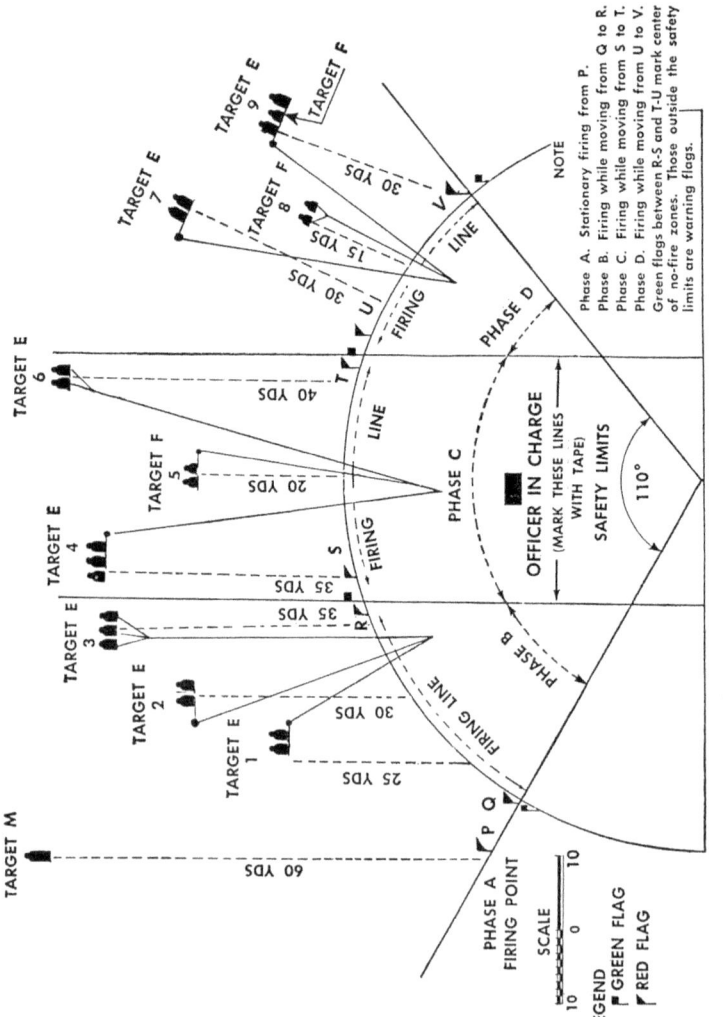

Figure 39. Dismounted practice range.

at his discretion. He will always expose group 3. As a result, the firer will fire at two groups of targets (five targets) in phase B.

(4) At the completion of phase B, the first firer will clear his weapon (checked by the scorer), move into the no-firing zone between phases B and C, and place his weapon on the stand. At the completion of phase A, the second firer will clear his weapon (checked by the scorer) and place it on the stand. On order of the officer in charge, both firers and their scorers move forward and score their respective targets.

c. *Phase C.*

(1) The first firer moves to a position on the firing line just short of the red flag designating the firing zone for phase C (point S, fig. 39). The second firer moves to the position at point Q. The third firer moves to the position at point P with four magazines loaded with 15 rounds in each.

(2) At the command of the officer in charge, the three firers take a crouched position; cock, lock, and load their guns; and announce READY. The officer in charge commands: FIRE WHEN READY.

(3) At this command, the firer at phase A (third firer) takes a standing position and commences firing phase A. The firer at point Q (second firer) walks along the phase B firing line and commences firing phase B. The firer at point S (first firer) walks along the phase C firing line and fires at group 4 and group 5 or 6, as the targets appear. The target operator will expose either group 5 or group 6, at his discretion. He will always expose group 4. As a result, the firer will fire at two groups of targets (five targets) in phase C.

(4) At the completion of firing, each firer will clear his weapon (checked by the scorer) and place it on the stand. On order of the officer in charge, the three firers and their scorers move forward and score their respective targets.

d. *Phase D.*

(1) The first firer moves to a position on the firing line just short of the red flag designating the firing zone for phase D (point U, fig. 39). The second firer moves to the position at point S. The third firer moves to the position at point Q. The fourth firer moves to the position at point P with four magazines loaded with 15 rounds in each.

(2) At the command of the officer in charge, the four firers take a crouched position; cock, lock, and load their guns; and announce READY. The officer in charge commands: FIRE WHEN READY.

(3) At this command, the firer at phase A (fourth firer) takes a standing position and commences firing phase A. The firer at point Q (third firer) walks along the phase B firing line and commences firing phase B. The firer at point S (second firer) walks along the phase C firing line and commences firing phase C. The firer at point U (first firer) walks along the phase D firing line and fires at group 9 and group 7 or 8, as the targets appear. The target operator will expose either group 7 or group 8, at his discretion. He will always expose group 9. As a result, the firer will fire at two groups of targets (five targets) in phase D.

(4) At the completion of firing, each firer will clear his weapon (checked by the scorer) and place it on the stand. On order of the officer in charge, the four firers and their scorers move forward and score their respective targets.

(5) After scoring is completed, the second, third, and fourth firers move to the next firing point, a new firer moves to point P with four magazines loaded with 15 rounds in each, and the first firer, who has completed the course, gives his score card to the recorder, behind the firing line.

e. Firing One Man at a Time. It is best, if time permits, to have only one man firing the course at a time. To do this, remove all flags and phase lines between point Q and point V. One scorer accompanies the firer throughout the course. After each firer begins phase B, he continues through phases B, C, and D without further commands, stopping only to reload.

f. Timer. Each target operator is responsible for exposing targets for the prescribed length of time.

g. Duties of Scorers.

(1) *Instruction practice.* During instruction practice, the scorers should also act as coaches and observe the action of the firers. At the completion of each phase, after the targets are marked and scored, the scorer should point out any errors made. Prior to, during, and after each phase, the scorer will require the firer to observe all safety precautions. Scorers are assigned to phases, not to firers; the firer has a different scorer at each phase.

(2) *Record firing.* During record firing, the scorer marks the targets and records the score on the firer's score card (fig. 47). HE WILL NOT COACH THE FIRER; COACHING IS NOT PERMITTED DURING RECORD FIRING. However, he will require the firer to observe all safety precautions.

h. Penalties. The first must take a crouched position, simulating cover, when loading the weapon. If he does not take this position, his targets should be immediately exposed, causing him to lose the opportunity of taking these targets under fire.

  i. *Scoring and Marking.*
   (1) *Procedure.* When the officer in charge has determined that the weapons are clear, he commands: MARK AND SCORE TARGETS. Firers and scorers mark or paste targets as indicated by the officer in charge.
   (2) *Scoring value.*
      (a) In phase A, the firer receives 5 points for each bullet hole in the target.
      (b) In phases B, C, and D, the firer receives 1 point for each target fired at. For example, if he fires at a group of two targets, he receives 2 points. If he fires at a group of three targets, he receives 3 points. Thus, if he fires at all groups in a phase, he receives 5 points.
      (c) In phases B, C, and D, the firer receives 5 points for each target that is hit, and 2 points for each bullet hole, not to exceed three hits on each target.
      (d) Maximum possible score for each phase is shown below.

Phase A—15 hits × 5 points per hit =	75
Phase B— 5 targets × 5 points (for each target that is hit) =	25
15 hits × 2 points (for each hit on target) =	30
5 targets × 1 point (for each target fired at) =	5
	60
Phase C—Same as phase B	60
Phase D—Same as phase B	60
POSSIBLE SCORE	255

      (e) Qualification scores are listed in paragraph 79.

  j. *Stoppage.* If a stoppage occurs, the firer will keep his weapon pointed down range and notify his scorer. The scorer will determine whether the firer or the weapon is at fault. If the firer is not at fault, the scorer will notify the officer in charge, who will order the phase fired again. If the firer is at fault, no refiring of the phase will be permitted, and the score obtained with the stoppage will stand.

83. Procedure for Familiarization Firing

  a. *General.* This course is designed to allow the men to become familiar with the weapon. It may be set up and fired in the following manner:
   (1) A straight firing line with several firing points is used, with

approximately 10 to 12 feet between firing points. Stands are provided on which to place the weapons during lulls in firing. Holders for type M targets should be placed at ranges of 25, 45, and 60 yards from the firing line. There is one target for each firing point.

(2) The M targets are first placed in the holders at the 25-yard range.

(3) Each firer has three magazines of 10 rounds each. Each firer has a coach who also acts as his scorer.

(4) The firers take the standing position. On the command of the officer in charge, they cock, lock, and load their guns. At the command COMMENCE FIRING, each firer fires four rounds, single shot, at his target. He then locks his weapon, takes the assault position, opens the cover, and fires a burst of six rounds at his target.

(5) The firers clear their weapons (checked by the coaches) and place them on the stands. On command, the firers and coaches move forward and mark and score the targets. They then move the targets to the 45-yard range, placing them in the holders there.

(6) After moving back to the firing line, the firers take standing, sitting, or kneeling positions. On command, they cock, lock, and load their guns and commence firing. Each firer fires four rounds, single shot, and two bursts of three rounds each, at his target.

(7) The firers clear their weapons (checked by the coaches) and place them on the stands. On command, the firers and coaches move forward and mark and score the targets. They then move the targets to the 60-yard range, placing them in the holders there.

(8) After moving back to the firing line, the firers take standing, sitting, or kneeling position. On command, they cock, lock, and load their guns and commence firing. Each firer fires four rounds, single shot, and two bursts of three rounds each, at his target.

(9) The firers clear their weapons (checked by the coaches) and place them on the stands. On command, the firers and coaches move forward and mark and score the targets. They then bring the targets back to the 25-yard range and place them in the holders there in readiness for the next order of firers.

(10) In some cases, a pistol range may be modified for firing this course by establishing the 45- and 60-yard ranges.

b. Duties of Coaches. Coaches perform the duties listed in paragraph 67. In addition, they score the targets and record the scores made by the firers.

c. *Scoring and Marking.*

(1) *Procedure.* When the officer in charge has determined that the weapons are clear, he commands: MARK AND SCORE TARGETS. Firers and coaches mark or paste targets as indicated by the officer in charge.

(2) *Scoring value.* Each hit on the target counts 5 points. Number of hits (30) x 5 points (each hit) = 150 points, the possible score.

(3) *Qualification scores.* There are no qualification scores for this course. In a unit, competition may be developed by establishing such scores.

Section V. TARGETS, RANGES, AND RANGE SAFETY PRECAUTIONS

84. Targets

a. Targets used in the dismounted practice course (fig. 39) are standard E, F, and M silhouettes (fig. 40). The target used for familiarization firing is the M silhouette.

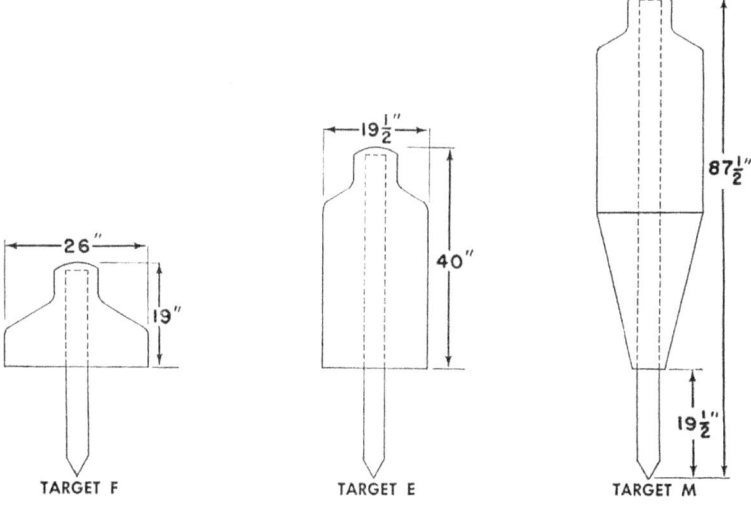

Figure 40. Targets.

b. Targets E and F are used in phases B, C, and D. Some are on pivots, some on hinges, and some on moving sleds or carriages.

(1) *Pivot type.* These are bobbing targets, located in groups 1, 2, 5, and 7. These targets are pulled upright for 2 seconds, then released to return to their defiladed positions behind the hill mask (figs. 41 and 42).

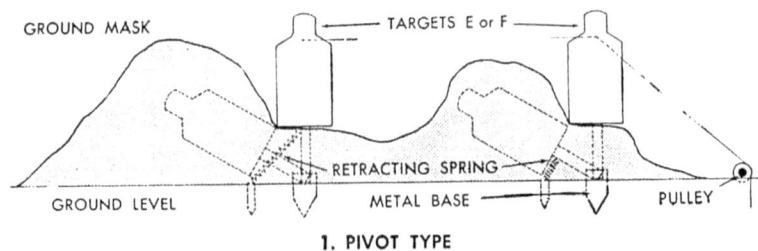
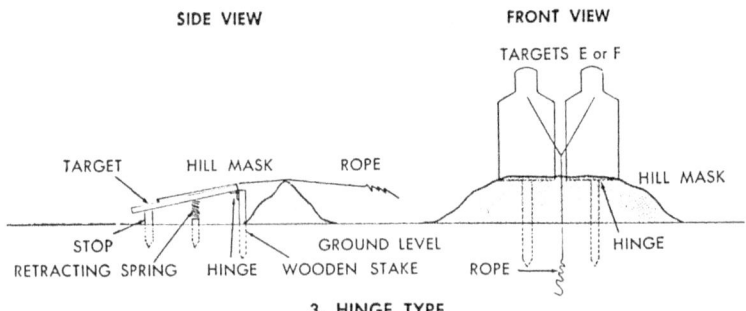

Figure 41. Methods of exposing stationary targets.

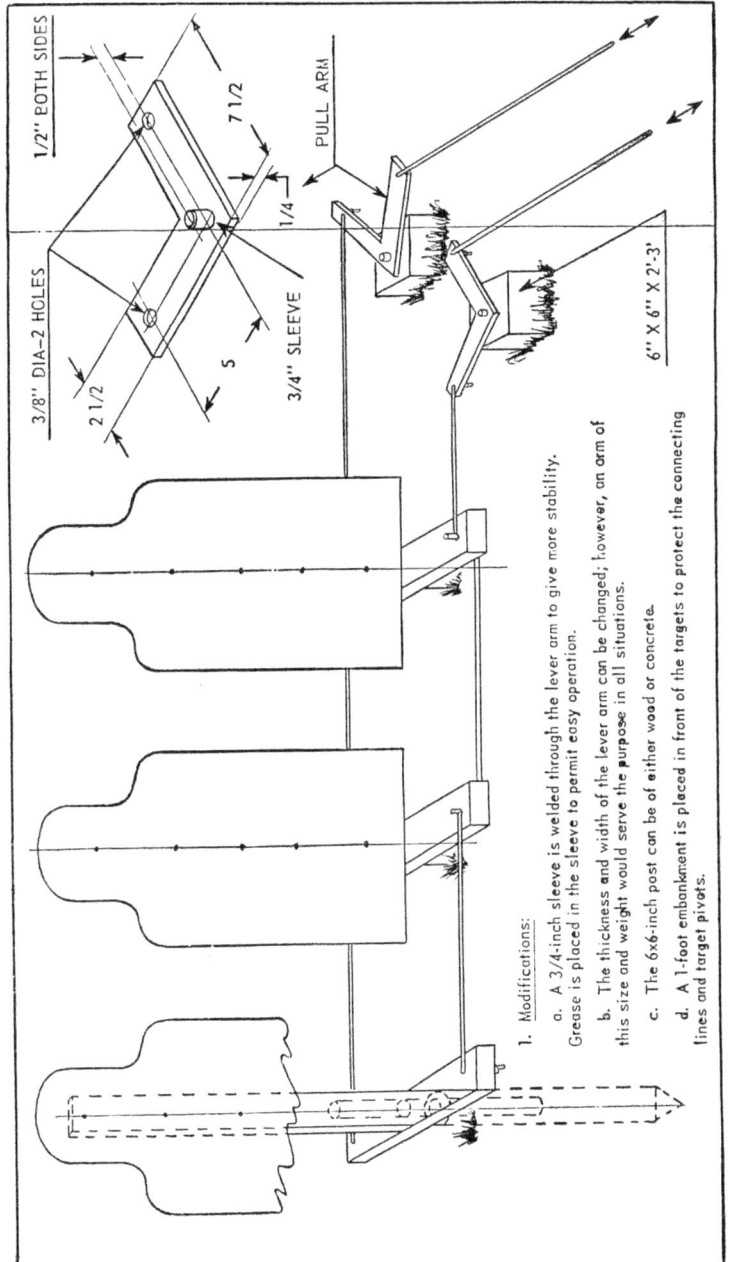

Figure 42. Alternate method of exposing stationary targets.

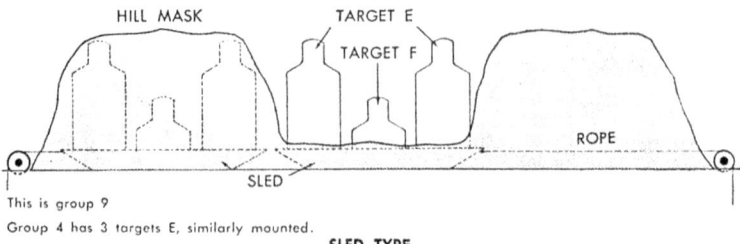

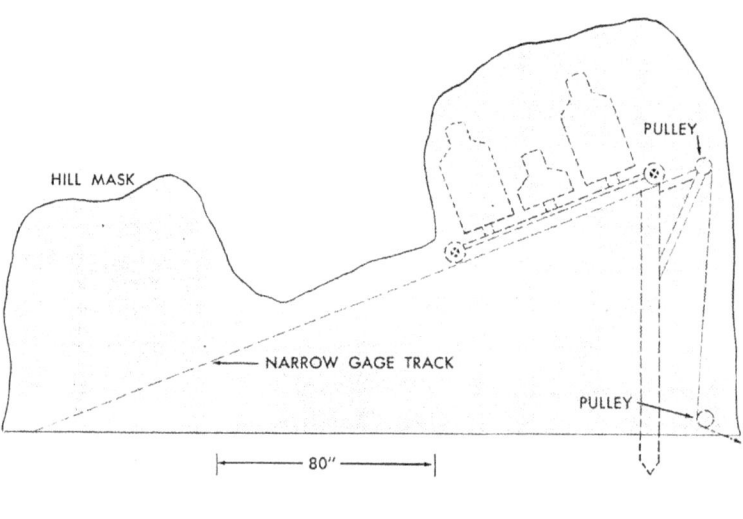

Figure 43. Methods of exposing moving targets.

(2) *Hinge type.* These are bobbing targets, located in groups 3, 6, and 8. These targets are pulled up and exposed for 2 seconds (figs. 41 and 42). The stop prevents the target from falling too far to the rear when released.

(3) *Sled or carriage type.* These are moving targets, located in groups 4 and 9. They are pulled or rolled across the opening (fig. 43) and are in view for 5 seconds while in motion.

85. Ranges

a. The dismounted practice course range (fig. 39) may be laid out on any ordinary flat terrain, preferably with some grass, weeds, and low underbrush. Targets may be partially concealed or may be near possible concealment in order to represent as nearly as possible actual enemy groups. If the available area is restricted in size, the firing line may be made straight to reduce the safety limits.

b. The familiarization course range may be set up on a pistol range (if space permits), a rifle range, or any flat terrain.

c. The moving ground target range (fig. 44) is used to familiarize the soldier with firing the submachine gun at a moving target (par. 95). Other types of moving target ranges may be used to fire this course.

d. The moving vehicle range (fig. 46) is used to familiarize the soldier with firing the submachine gun from a moving vehicle at ground targets.

86. Range Safety Precautions

During firing, all personnel, including marking and scoring details, must be in a safe position. The necessary range guards must be posted and danger flags prominently displayed before firing begins. Men on the firing line must observe all safety precautions.

CHAPTER 5

MARKSMANSHIP, MOVING GROUND TARGETS AND VEHICULAR FIRING

Section I. FIRING AT MOVING GROUND TARGETS

87. General

a. All personnel armed with the submachine gun will be trained to fire at moving vehicular and personnel-type targets. Normally such fire will be delivered at short ranges in short bursts. The high rate of fire, and the ability of the soldier to move the direction of fire at will, make the submachine gun particularly effective against moving personnel. The soldier must be trained to employ his submachine gun effectively and quickly. He must be trained in the proper use of the sights and methods of leading the target at short ranges.

b. Moving targets are seldom exposed for long periods and usually move at maximum speed during periods of exposure.

c. Firing at moving targets with service ammunition should follow firing of the dismounted practice course and record firing. The moving ground target course is not included in record practice.

88. Use of Leads

When targets are crossing the line of sight, the firer must aim ahead of the target so that the bullet and target will meet. The distance aimed ahead of the target is called the *lead*. For personnel targets moving across the line of sight, the point of aim should be slightly in front of the body, and the lead should be corrected by observation of the fire. Targets which approach directly toward the firer or move directly away from him require no lead.

89. Determination of Leads

The lead necessary to hit a moving target depends upon the speed of the target, the range to the target, and the direction of movement with respect to the line of sight. Moving at 10 miles per hour, a vehicle travels approximately its own length in 1 second. The velocity of a bullet from the submachine gun is approximately 900 feet or 300 yards in 1 second.

90. Application of Leads

a. Leads are applied by using the length of the target (TL), as it appears to the firer, as a unit of measure. This eliminates the necessity for corrections due to the angle at which the target crosses the line of sight; because the more acute the angle, the shorter the target appears, and the less lateral speed it attains.

b. The following lead table for vehicles is furnished as a guide:

Miles per hour	Range
	100 yards or less
10	1/3 TL
20	2/3 TL

91. Technique of Fire at Moving Targets

a. The firer uses the following technique in firing at moving targets:

(1) *Approaching or receding targets.* The firer holds his aim on or above the center of the target (depending upon the range) and fires in short bursts.

(2) *Crossing vehicular targets.* The firer estimates the proper amount of lead, aligns his sights on or above the bottom of the target at its rearward point (depending upon the range), swings straight across the target to the estimated lead, and fires short bursts, keeping the proper lead.

(3) *Crossing personnel targets.* The firer takes aim slightly in front of the center of the body of the target, with proper adjustment for range, and fires short bursts. He changes the lead and range as necessary after observing the effect of the bursts.

b. The high rate of fire of the submachine gun allows the firer to cover the target with fire and to improve his lead and range estimation by actual observation of the effectiveness of his fire.

Section II. MOVING TARGET RANGE AND SAFETY PRECAUTIONS

92. General

The ability of the soldier to hit a target moving on the ground is developed through exercises conducted as a part of the combat firing of his organization. The following course is an exercise which is used to obtain this result. In the moving target course, the soldier fires at a moving ground target from a vehicle that is halted when he fires but which moves between bursts.

93. Moving Targets

The target consists of rectangular frame 5 feet high by 8 feet in length, covered with target cloth or other light-colored material, and mounted on a carriage which can move at a speed of 20 miles per hour. The substitution of E and M targets for the target cloth will provide a suitable target for training in firing at moving personnel.

94. Moving Target Range Construction

a. A moving ground target range is shown in figure 44. A, B, and C represent parapets of sufficient size to completely hide the target from the firer's view, with dugouts in the rear for protection of the pit details and scorer. Y is the starting point of the firer's vehicle. The route for the firer's vehicle (Y to B) should be smooth and level. The entire area from Y to the parapets should be clear in order that ground firing points may be employed for firing at moving personnel targets.

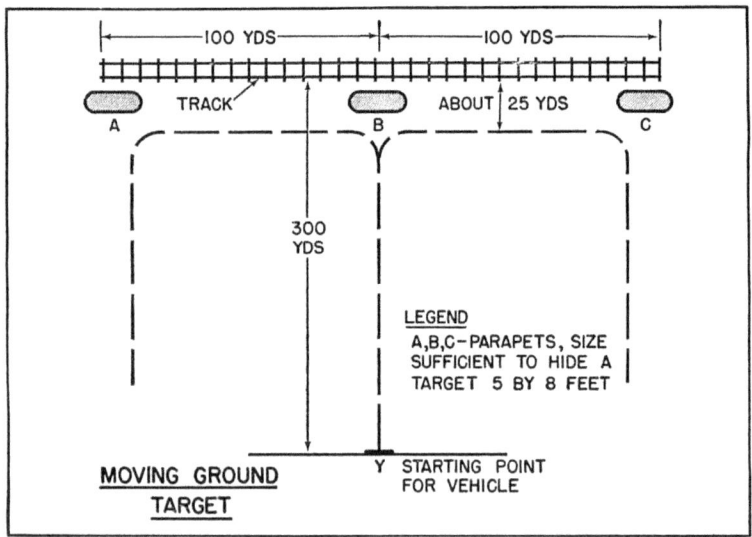

Figure 44. Moving ground target range.

b. The dotted line in figure 45 represents a steel cable which is fastened to the target carriage. The cable is run from A, along the top of the track, to and around a pulley, B; then back under the target track and carriage to a cylindrical drum, C, around which it goes twice; then around a pulley, D, which is mounted on a movable frame to adjust tension in the cable; then back to the target carriage at E. The shaft (or axle) of drum C is attached through a transmission and clutch to a motor. Thus the drum may be rotated in one direction by running the motor with the transmission in reverse and in the other

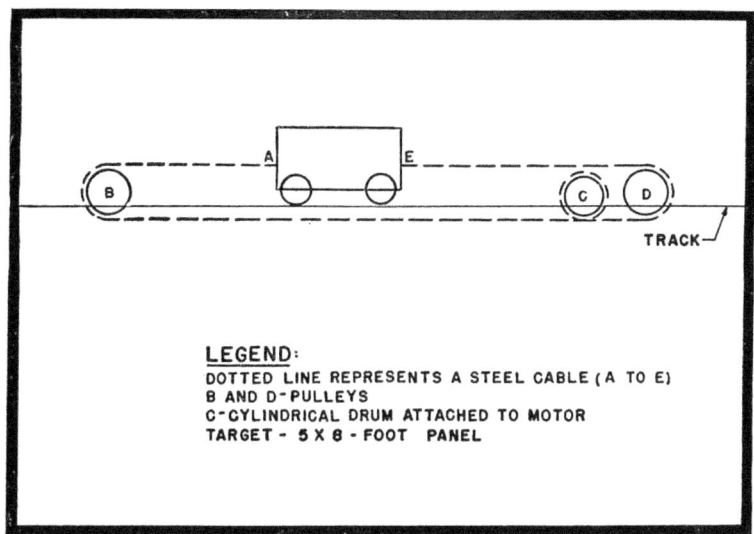

Figure 45. Moving target arrangement.

direction with the transmission in a forward speed. The target carriage is equipped with flanged wheels and is run on narrow-gage tracks.

c. When the necessary material is not available for the construction and operation of such an installation, a simpler arrangement can be made by towing a double-ended sled behind a vehicle. However, with such a sled, the target cannot be stopped and started again in either direction as easily.

d. Firing points for firing at moving personnel are established at ranges less than 100 yards from the target.

95. Moving Target Course and Firing Procedure

a. *Procedure.* The firer starts at Y (fig. 44). He is mounted in a vehicle, with his weapon cocked, locked, and loaded with a magazine of 20 rounds. The vehicle moves forward toward B and attains a speed of 15 to 20 miles per hour. The target may be behind either parapet A or parapet C. Upon telephone or visual signal from behind the starting line, the target is released and moves at a speed of 10 to 20 miles per hour toward parapet B. When the target appears, the vehicle is stopped. The firer unlocks his gun and, from the stationary vehicle, fires at the moving target as long as it is visible. The target should be released when the range from the firer to the target is not more than 100 nor less than 75 yards. The firer fires all 20 rounds in his magazine in bursts of about three rounds each, fired in rapid succession. When the target disappears behind parapet B, the firer quickly reloads with a new magazine of 20 rounds and calls READY; then the vehicle again moves forward at a speed of 10 to 20 miles per hour. Meanwhile, the target is held behind parapet B for 10 seconds and is then released; it may move toward either parapet A or parapet C. When the target appears, the vehicle is stopped. The firer unlocks his gun and, from the stationary vehicle, fires at the moving target as long as it is visible. He fires all 20 rounds in his magazine; the method of delivering fire is optional.

b. Scoring. The course is scored as follows:
> 20 points for hitting the target.
> 2 points for each hit on the target.
> Maximum possible score is 100 points.

c. Coaches.
   (1) A coach should be designated to accompany each firer. At the completion of firing, he corrects any errors made.
   (2) Prior to, during, and after firing the coach will require the firer to observe all safety precautions.
   (3) The coach will insure that a red flag is displayed from the vehicle whenever the weapon is loaded. When the weapon is clear, a green flag will be displayed.

d. Stoppage. In the event of a stoppage, the target is halted and the stoppage corrected; then firing is resumed.

96. Moving Target Range Safety Precautions

A red flag will be displayed at all times to indicate that the range is in use. A red flag will be displayed from the vehicle while the gun is loaded and during firing. Any necessary range guards will be placed out. At the pits, a red flag is displayed at any time the pit detail personnel are not under cover. The officer in charge of the pit detail will take the red flag down when all members are under cover, indicating that it is safe to fire. Telephone communication with the pits is desirable. The right and left limits of fire must be plainly marked by posts or flags.

Section III. VEHICULAR FIRING (OPEN VEHICLE)

97. General

a. Personnel armed with the submachine gun are trained to fire from stationary and moving vehicles at appropriate targets if time and ammunition allowances permit. Practice in firing from moving vehicles follows instruction in firing the dismounted practice course and record firing.

b. The course outlined in this section is furnished as a guide. Direction of fire, length and contour of the track, arrangement of the targets, and ranges to the targets may be varied to suit local conditions.

The three phases outlined may be fired separately. It is desirable that all personnel at an installation fire the same course. This course is not included in record practice.

98. Vehicular Range Construction

a. A range for firing from an open vehicle is shown in figure 46.

b. Targets are stationary E and F silhouettes, arranged in groups.

c. The range may be located on any ordinary terrain that has some grass, weeds, and underbrush and a good field of fire for the firer. Targets may be partially concealed but must be easily located by the firer. The road for the vehicle should be smooth, particularly along those portions where firing occurs.

d. The limits of the safety area (AR 385-63) may be modified by local authorities, with the approval of higher authority, when the nature of the terrain or artificial barriers make a smaller area safe.

99. Vehicular Firing Course

Two magazines loaded with 25 rounds in each are allowed for firing the course. Each man will fire the course twice. Additional firing may be done if ammunition allowances permit.

100. Procedure for Firing the Vehicular Course

The firer takes his place in the front of the vehicle, which is about 50 yards from point A. The firer carries with him two magazines loaded with 25 rounds in each.

a. Phase A. At the command of the officer in charge of firing, the firer cocks, locks, and loads his weapon, and calls READY. The officer in charge directs the vehicle to start on the course. The vehicle reaches and maintains a uniform speed of approximately 20 miles per hour. The firer may start firing on the targets in group 1 as soon as the vehicle passes point A. The firer remains low in the vehicle and exposes himself only when firing. He fires approximately 15 rounds at the targets in group 1. He must cease firing when the vehicle reaches point B, which is 75 yards from point A.

b. Phase B. The vehicle continues on the course. When it reaches point C, the firer may commence firing on the targets in group 2. He fires all of the remaining rounds from the first magazine at this group. He must cease firing when the vehicle passes point D.

c. Phase C. After firing at group 2, the firer moves to the rear of the vehicle or to some other position from which he can fire to the rear. He loads the second magazine into his weapon. The vehicle continues on the course until it reaches point E, where it halts for 15 seconds. The firer can commence firing on the targets in group 3 as

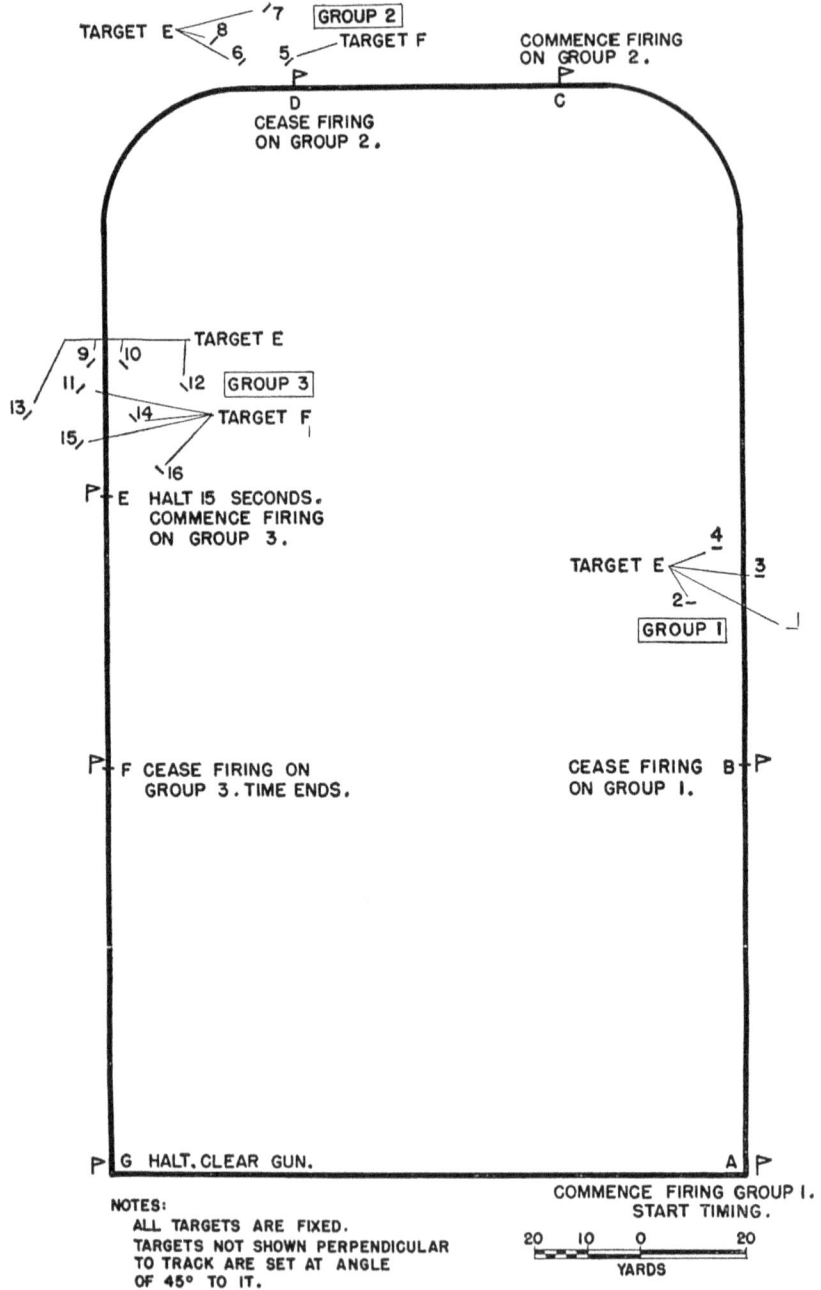

Figure 46. Moving vehicle range.

soon as the vehicle has come to a full halt. He may continue to fire after the vehicle resumes moving on the course. He must cease fire when the vehicle reaches point F. Upon completion of the course at point G, the vehicle is halted, the firer clears his weapon (checked by the coach), the coach displays the green flag, and the vehicle moves to the starting position.

  d. *Scoring and Marking.*

   (1) At completion of firing and on order of the officer in charge of firing, the scorer moves out in a vehicle and proceeds around the course. He records the firer's score, and marks or pastes the targets as indicated by the officer in charge.

   (2) Targets are scored as follows:

16 targets x 5 points (for each target that is hit)	= 80
45 hits x 2 points (for each hit on targets 1 through 15, except that not more than three hits per target will be counted)	= 90
5 hits x 2 points (for each hit on target 16; not more than five hits on this target will be counted)	= 10
POSSIBLE SCORE	180

  e. *Duties of the Coach.* The coach takes a position in the vehicle so as not to interfere with the firer. The coach may assist the firer to commence and cease fire at the designated points in the course by tapping the firer lightly on the back or shoulder. He will command CEASE FIRING at the proper points during the run, and will tell the driver when to halt and when to move. At point E in phase C, the coach will time the halt by the vehicle. At the end of 15 seconds, he will direct the driver to start the vehicle moving down the course. Prior to, during, and after firing he will require the firer to observe all safety precautions. He will display a red flag from the vehicle when the weapon is loaded, and a green flag when the weapon is clear.

  f. *Stoppage.* In the event of a stoppage, the vehicle is halted and the stoppage corrected; then firing is resumed.

CHAPTER 6

TECHNIQUE OF FIRE AND DESTRUCTION OF MATERIEL

Section I. TECHNIQUE OF FIRE

101. Characteristics of Fire

a. General. The characteristics of the submachine gun govern the manner in which it is used. It is highly effective at close quarters. It is very dependable, because of the simplicity of its mechanism. The soldier who uses the weapon properly can attain considerable accuracy in firing automatic fire at close ranges.

b. Collective Firing. Collective firing is the combined firing of a group of individuals. The submachine gun is normally issued as an individual weapon. It is not issued to all members of a unit, as is the pistol, carbine, or rifle. Consequently, collective firing of submachine guns is seldom employed. The submachine gun may be used in conjunction with other weapons, especially the machine gun. When used in this manner, it is normally fired at short-range targets, while the other weapon is fired at longer ranges.

102. Types of Fire Commands

a. Formal fire commands are seldom necessary or desirable. For control on target ranges, such commands as COCK, LOCK, AND LOAD; READY; COMMENCE FIRING; and CEASE FIRING are used. In combat, fire commands, if necessary, are normally limited to COMMENCE FIRING and CEASE FIRING.

b. The target designation may be added to the fire command when the firer has not seen the target. In this case, the fire command may be given as follows: JONES, RIFLEMAN BEHIND TREE TO THE RIGHT, COMMENCE FIRING. Normally, each soldier discovers targets and immediately takes them under fire.

103. Sample Technique-of-Fire Exercises

In preparing exercises involving the use of the submachine gun, advantage is taken of field exercises and maneuvers to present logical situations, some phases of which would require the employment of this weapon both from the ground and from a vehicle. These exercises should include the use of the submachine gun in the dismounted reconnaissance of a roadblock, its employment on outpost duty or in establishing march outposts, and its use by mounted scouts and armored vehicle personnel in assumed ambush situations.

a. Exercises, General. The following exercises are given as a guide and may be modified to suit the terrain, equipment, and time available. Each problem utilizes natural terrain features, equipment normally available, and actual personnel targets (all fire is simulated). The exercises should be conducted under the supervision of a commissioned officer, who will note any errors made and critique each exercise. Service ammunition is not fired during these exercises. They train the soldier and unit leader in locating targets, target designation, fire commands, and the use of the submachine gun as a supporting weapon in the unit. Every effort is made to employ the fundamentals of concealment, camouflage, and scouting and patrolling in the conduct of these exercises. Personnel acting as targets should be rotated with firers, and the targets should be shifted frequently to avoid monotony.

b. Exercise 1. A stretch of terrain not to exceed 400 yards in length and containing as many terrain features as possible—such as trees, shrubs, tall grass, ditches, logs, and walls—is selected for the course. Actual personnel targets are placed along a designated path, at various ranges from the path and in normal concealment. Typical targets include prone, kneeling, and standing soldiers, moving individuals and groups, machine guns with normal crews, and mounted scouts. The firer is required to proceed down the designated path and locate targets. When he discovers a target, he takes a firing position and simulates firing on the target. He is accompanied by an instructor, who checks all phases of the firer's actions and points out any errors made.

c. Exercise 2. A roadblock is established in a suitable location. It is held by a detachment of soldiers armed with submachine guns and rifles. Either a dismounted soldier with submachine gun, a mounted scout with submachine gun, an armored vehicle with submachine gun as an alternate weapon, or any combination of these, may operate against the roadblock. A commissioned officer should accompany the individual or the vehicle. He checks and instructs in procedure and critiques the exercise, including any commands given by a vehicle commander.

d. Exercise 3. A small area in which buildings predominate, and which can be presumed to be a village or city street, should be selected for this exercise. It should be possible for personnel to occupy buildings, roofs, and windows and to erect barricades. Personnel armed with submachine guns and mounted in armored vehicles should be required to operate against personnel in buildings and to reduce barricades. This type of problem is especially beneficial in training for mounted and dismounted action, collective firing with other weapons, and proper leadership. All actions by individuals, squad or platoon leaders, and units should be carefully checked by a commissioned officer. The exercise should be reviewed and critiqued immediately upon completion.

Section II. DESTRUCTION OF MATERIEL

104. Authority

The decision to destroy ordnance materiel to prevent its capture and use by the enemy is a command decision. Only a division or higher commander has authority to order such destruction.

105. Principles Governing Destruction

The following are the fundamentals to be observed in executing an order to destroy small arms.

a. The destruction must be as complete as the circumstances will permit.

b. If there is insufficient time for complete destruction, the parts essential to operation of the weapon must be destroyed, beginning with those parts most difficult for the enemy to duplicate.

c. The same essential parts of each weapon must be destroyed, to prevent the reconstruction of a complete weapon from several damaged ones.

106. Training in Destruction

Before reaching the combat zone, soldiers must be trained to quickly and adequately destroy their individual weapons in an established and uniform sequence, based on the principles stated in paragraph 105. Training will not involve actual destruction of materiel.

107. Method for the Destruction of the Submachine Gun

Remove and dispose of the bolt and guide rod group. Smash the receiver and stock against a tree, rock, or other hard surface until bent and twisted. Smash or burr the threads of the barrel collar.

108. Destruction of Ammunition

When time and materials are available, ammunition may be destroyed as follows: Break out all packed ammunition from boxes or cartons. Stack the ammunition in a pile. (If possible, the pile should be placed in a depression or hole, to lessen the danger to personnel performing the destruction operation.) Stack or pile all available inflammable material, such as scrap wood or brush, over the ammunition. Pour gasoline or oil over the pile. Sufficient inflammable material must be used to insure a very hot fire. Ignite the material and take cover. A period of 30 to 60 minutes will be required to destroy the ammunition carried by small combat units.

CHAPTER 7

ADVICE TO INSTRUCTORS

Section I. GENERAL

109. Purpose

The material contained in this chapter is advisory and should be considered as a guide only. It is not intended to limit the imagination and initiative of the instructor.

110. Assistant Instructors

Train, in advance of classes, as many demonstrators and assistant instructors as will be needed. Rehearse them carefully in the duties they are to perform. Only by rehearsals can the instructor insure effective demonstrations and efficient work by assistant instructors.

111. Training Schedules

To aid in the individual training phase, a training schedule for a course in marksmanship training is shown in paragraphs 114 and 115. This schedule is based on the desirable number of training hours for a submachine gun course. Use it as a guide in preparing lesson plans. Conditions may require a longer or shorter period to complete the training. When time is available, more training should be added to the schedule. When suggested references, equipment, and training aids are not available, improvise or substitute the best that are available. A familiarization training schedule with training notes is included in this chapter. All references in the training schedule, unless otherwise noted, are to this manual.

112. Allotment of Training Hours for the Marksmanship Course

The suggested allotment of time for conducting this training and firing is based on the requirement of a company-size unit, not on the amount of time spent on each element by the individual soldier.

Subject	Hours
Mechanical training	5
Preparatory marksmanship training	13
Range firing.	
a. Instruction, dismounted practice course	9
b. Record, dismounted practice course	3
c. Instruction, moving ground targets	8
d. Instruction, vehicular firing	12
Total	50

113. Allotment of Training Hours for the Familiarization Course

The total number of hours for this training may be reduced if the personnel being instructed have had previous training on the submachine gun.

Subject	Hours
Mechanical training	2
Preparatory marksmanship training	3
Range firing (two orders)	2
Total	7

Section II. TRAINING SCHEDULE AND TRAINING NOTES FOR SUBMACHINE GUN MARKSMANSHIP COURSE

114. Mechanical Training (5 Hours)

P[1]	H[2]	Lessons	Text references	Area	Training aids and equipment
1	1	Characteristics, general data, nomenclature, field disassembly and assembly.	Par. 3–12.	Preferably a large classroom with tables and chairs.	For instructor: submachine gun, dummy cartridges, blackboard, and GTA 9–3–1. For student: submachine gun, dummy cartridges, and disassembly mat GTA 9–618.
2	1	Detailed disassembly and assembly, operation check, ammunition, and filling magazines.	Par. 13–15, 28, 29, 49–55.	Do.	Same as for period 1.
3	1	Functioning.	Par. 16–26.	Do.	For instructor: same as for period 1 plus working model (fig. 48) and overhead projector. For student: same as for period 1.
4	1	Malfunctions, stoppages, and immediate action; care and cleaning.	Par. 33–47.	Do.	For instructor: same as for period 1 plus cleaning materials. For student: same as for period 1 plus cleaning materials.

[1] P—Period.
[2] H—Hours.

P¹	H²	Lessons	Text references	Area	Training aids and equipment
5	1	Examination.	All previous references.	Do.	For student: submachine gun, pencil and paper.

115. Preparatory Marksmanship Training (13 Hours)

P¹	H²	Lessons	Text references	Area	Training aids and equipment
6	2	Orientation on marksmanship; sighting and aiming exercises.	Par. 63–71.	Drill field or other suitable training area.	For instructor: submachine gun, submachine gun rest, sighting disk, sighting board, and blackboard. For student: same as instructor less blackboard, plus pencil and paper.
7	2	Firing positions: standing, sitting, kneeling, prone, and assault.	Par. 72.	Do.	For instructor: submachine gun. For student: submachine gun.
8	1	Trigger manipulation, safety precautions, loading and unloading.	Par. 29–32, 73, 129–132.	Do.	For instructor: submachine gun and blackboard. For student: submachine gun.
9	6	Marksmanship exercises.	Par. 74.	Dismounted practice range.	All range equipment.
10	2	Examination.	Par. 75, 76; all previous references.	Preferably a large classroom with tables and chairs.	For student: submachine gun, pencil and paper.
11		Range firing: Time required depends upon the number of men to fire.			

¹ P—Period.
² H—Hours.

116. Training Notes, Mechanical Training

a. Instruction in mechanical training will be conducted in a sequence that insures the uniform progress of the unit.

b. The instructor briefly explains the subject to be covered. The assistants demonstrate the proper procedure for clearing the weapon. The students then clear their weapons. The instructor names the parts, and the assistants point out each part as it is named. The assistants demonstrate disassembly and assembly and then supervise the students during practical work on disassembly and assembly.

c. The instructor teaches functioning, stoppages, and malfunctions with the use of visual aids.

117. Training Notes, Preparatory Marksmanship Training

a. The conference at the beginning of each step is an important part of this training. Under no circumstances should the instructor read the material to the class. The important thing is to show the students, by explanation and demonstration, how to go through the exercises and to impress upon them why the exercises are given.

b. During the marksmanship exercises, coaches critique each exercise and help the students during the exercises.

c. The examination should be both practical and written. Extra instruction should be given to those students who are weak in any of the subjects.

Section III. TRAINING SCHEDULE AND TRAINING NOTES FOR SUBMACHINE GUN FAMILIARIZATION COURSE

118. Mechanical Training (2 Hours)

P[1]	H[2]	Lessons	Text references	Area	Training aids and equipment
1	1	Characteristics, general data, nomenclature, field disassembly and assembly.	Par. 3–12.	Preferably a large classroom with tables and chairs.	For instructor: submachine gun, dummy cartridges, blackboard, and GTA 9–3–1. For student: submachine gun, dummy cartridges, and disassembly mat GTA 9–618.
2	1	Functioning, malfunctions, stoppages, immediate action, and care and cleaning.	Par. 16–26, 33–47.	Do.	For instructor: same as period 1 plus working model (fig. 48), overhead projector, and cleaning materials. For student: same as period 1 plus cleaning materials.

[1] P—Period.
[2] H—Hours.

119. Preparatory Marksmanship Training (3 Hours)

P¹	H²	Lessons	Text references	Area	Training aids and equipment
3	1	Orientation on marksmanship, sighting and aiming exercise, and trigger manipulation.	Par. 63–71, 73.	Drill field or other suitable training area.	For instructor: submachine gun, submachine gun rest, sighting disk, sighting board, and blackboard. For student: same as instructor less blackboard, plus pencil and paper.
4	1	Firing positions and safety precautions.	Par. 32, 72, 129–132.	Do.	For instructor: submachine gun. For student: submachine gun.
5	1	Examination.	Par. 75, 76; all previous references.	Preferably a large classroom with tables and chairs.	For student: submachine gun, pencil and paper.
6		Range firing: Time required depends upon the number of men to fire.			

¹ P—Period.
² H—Hours.

120. Training Notes, Familiarization Course

a. The purpose of the familiarization course is to give the soldier enough training to enable him to maintain and fire the submachine gun during an emergency.

b. If the students have received training on other weapons, the time spent on sighting and aiming may be reduced.

Section IV. RANGE FIRING

121. Inspection of Submachine Gun

All submachine guns should be carefully inspected far enough in advance to permit the replacement of defective weapons before the training period begins.

122. Ammunition for Range Firing

The best ammunition available is reserved for record firing. Ammunition of different lot numbers should not be mixed.

123. Vehicles and Drivers for Vehicular Firing

The best vehicles and drivers of each organization should be made available for vehicular firing. Vehicles must be suitable for the type firing desired.

124. Inspection of Ranges

All ranges to be used are carefully inspected far enough in advance of the period of use to permit changes or repairs when necessary. Targets and other equipment must be in the best possible state of repair when range firing begins.

Section V. SCORE CARD

125. General

A sample individual score card is shown in figure 47, provisions of paragraph 20, AR 310-1 apply.

INDIVIDUAL SCORE CARD

SUBMACHINE GUN   QUALIFICATION. . . . .
QUALIFICATION COURSE   POSSIBLE. 255
NAME_____GRADE_____   EXPERT. 180
ORGANIZATION_____DATE_____   SHARPSHOOTER. 160
                                          MARKSMAN. 140

			HITS	TOTAL
PHASE A--SLOW FIRE 5 rounds standing 5 rounds kneeling 5 rounds prone (automatic fire) POSSIBLE 75 POINTS			X5 =	
COMBAT TARGETS--ANY METHOD OF FIRE OR POSITION		GROUPS ENGAGED	TARGETS HIT	
PHASE B Either group 1 or group 2 will be exposed; group 3 must be exposed. POSSIBLE 60 POINTS	Group 1	X2 =	X5 =	X2 =
	Group 2	X2 =	X5 =	X2 =
	Group 3	X3 =	X5 =	X2 =
PHASE C Either group 5 or group 6 will be exposed; group 4 must be exposed. POSSIBLE 60 POINTS	Group 4	X3 =	X5 =	X2 =
	Group 5	X2 =	X5 =	X2 =
	Group 6	X2 =	X5 =	X2 =
PHASE D Either group 7 or group 8 will be exposed; group 9 must be exposed. POSSIBLE 60 POINTS	Group 7	X2 =	X5 =	X2 =
	Group 8	X2 =	X5 =	X2 =
	Group 9	X3 =	X5 =	X2 =

NOTE: In phases B, C, and D, do not count more than three hits on a single target.

TOTAL_____

Signature of Firer   Signature of Witnessing Officer   Grade   Organization

Figure 47. Sample individual score card.

Section VI. TRAINING AIDS
126. General

In conducting training, make the greatest possible use of working models, charts, and other suitable visual training aids. Excellent submachine gun training aids are available through normal supply channels, or they can be constructed locally. Some recommendations and suggestions concerning the construction and use of the various training aids are listed below.

a. Scrap lumber in good condition may be used to construct models. Hardwood is recommended for model parts that rub together or have strain on them. Other materials which are needed in making models are nails, screws, and springs.

b. Charts and models should be painted with contrasting colors to help the class locate the various parts.

c. Models should be mounted on stands so that they can be seen by the entire class.

d. All lettering on charts, models, and other training aids must be large enough to be read easily. Lettering 2½ inches high is readable at a distance of 75 feet.

e. Charts which are painted or drawn on heavy paper will be just as satisfactory as those painted on wood, except that charts painted on wood will withstand harder usage than paper charts.

f. Before using a multicolored chart or working model, explain the meaning of the various colors.

127. Wooden Working Model

A large-scale working model (fig. 48) is an excellent training aid for teaching functioning of the submachine gun. Figures 48 and 49 show the dimensions of the parts, and a detailed plan of construction is given below.

a. Cut out the parts.

b. Paint the outline of the receiver on a sheet of ½-inch plywood.

c. Position the rails for the barrel (B), place the cartridge between the rails, and fasten the rails to the receiver with screws.

d. Position the guide rod (C). The bottom of the guide rod should be 4⅜ inches below the top of the receiver. Fasten it to the receiver with screws.

e. Slide the bolt forward on the guide rod.

f. Position the guide rod retaining plate (D), and fasten it to the receiver with screws.

g. Position the magazine follower (E) by placing the cartridge on

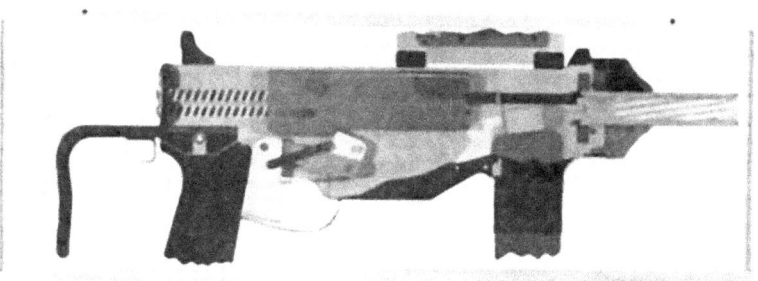

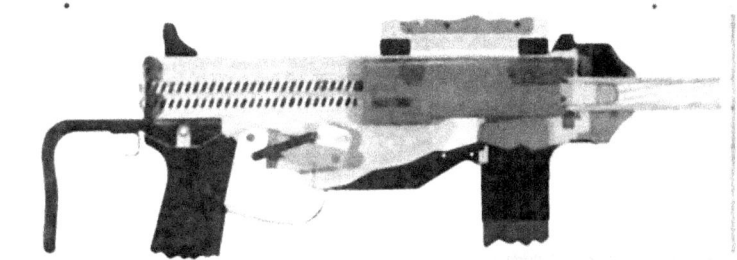

Figure 48. A wooden working model.

top of the follower. Move the bolt forward. The bolt will push the cartridge forward; guide it (by hand) into the chamber. Mark a path for the lug of the cartridge. Recess the path of the lug (A) to a depth of ¼ inch. When the model will feed and chamber the cartridge, fasten the follower to the receiver.

h. Assemble the trigger, connector, and sear (F), using a 1-inch dowel rod.

i. Position the trigger and sear group on the receiver.

j. Place the trigger pin in position, and fasten it to the receiver with screws.

k. Place the sear pin in position, and fasten it to the receiver with screws.

l. Place on the trigger spring (use a salvage driving spring).

m. Chamber the cartridge, and place stops in the barrel to prevent the cartridge from going completely into the barrel.

n. Paint the ejector on the receiver.

128. Charts and Other Aids

a. Some of the points that can best be presented by a chart are—

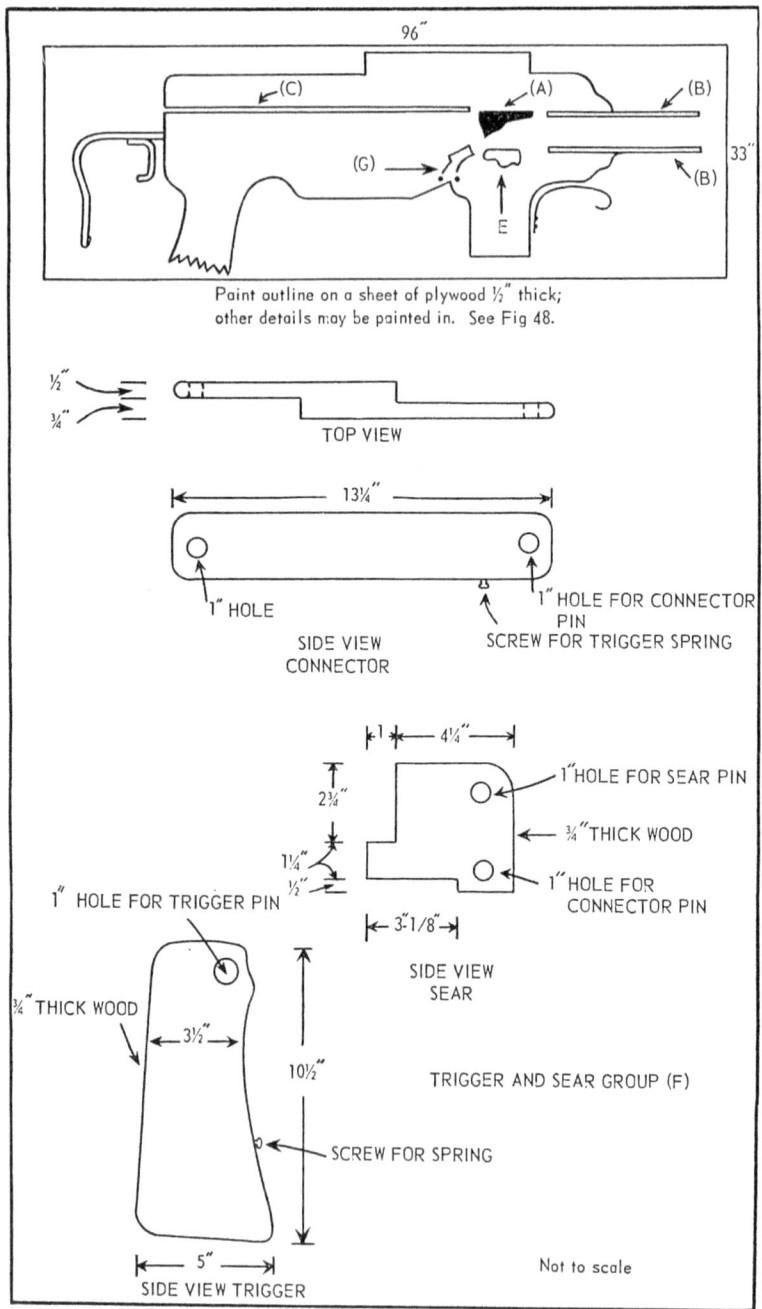

Figure 49. Diagram of working model.

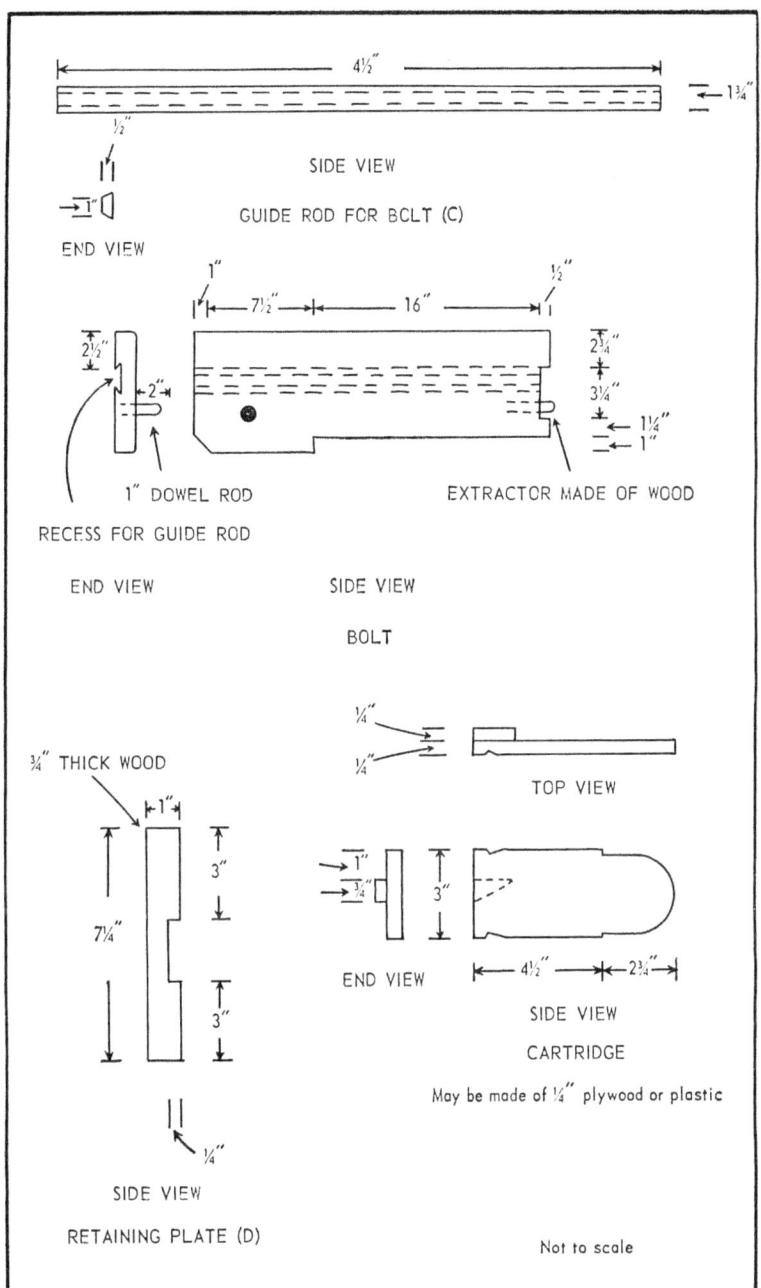

Figure 49—Continued.

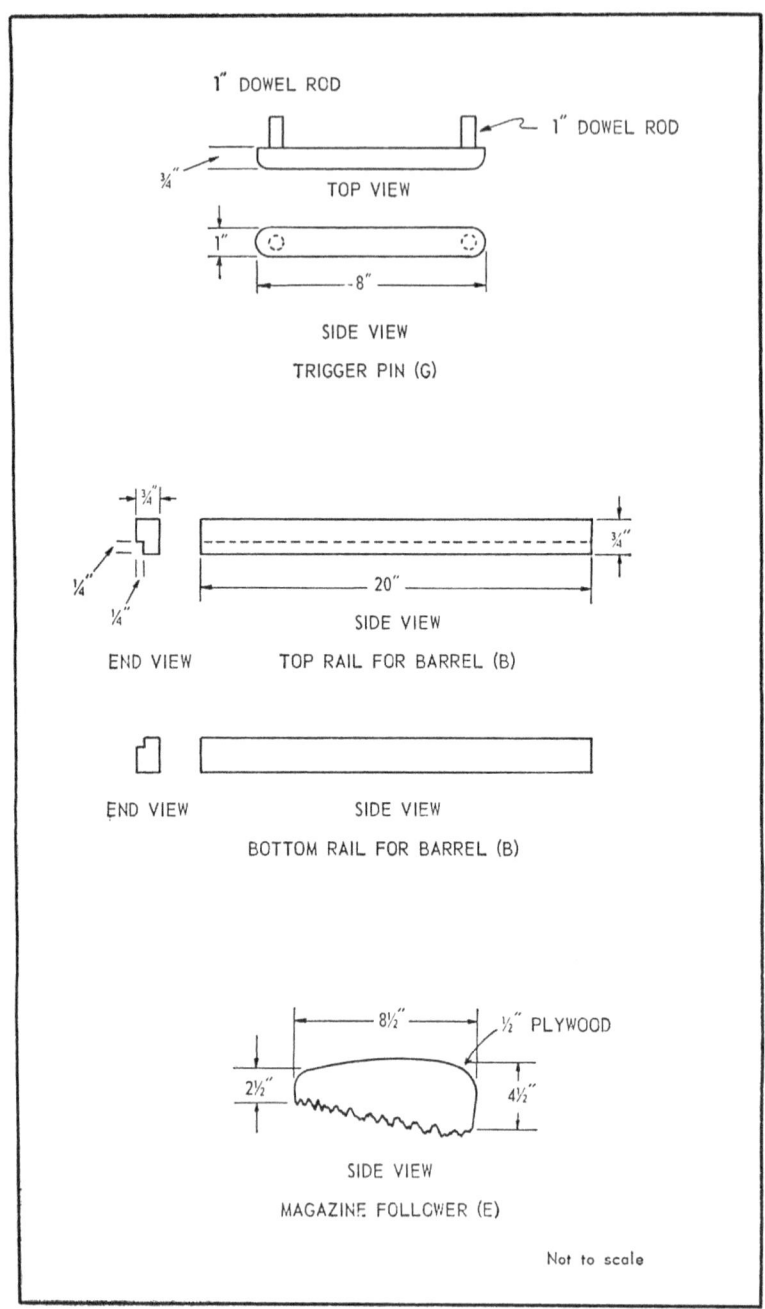

Figure 49—Continued.

(1) Steps in functioning.
(2) Stoppages.
(3) Steps in care and cleaning.
(4) Characteristics and general data.
(5) Range orientation.

  b. Other helpful training aids which can be obtained through normal supply channels are—
(1) Training films and film strips.
(2) Graphic training aids. These include—
  (*a*) GTA 9-3-1, available in 2x2-inch slides for use with slide projector, in 8x10-inch transparencies for use with the overhead projector, and in chart form.
  (*b*) GTA 9-6-2, available in an 8x10-inch transparency.
  (*c*) GTA 9-618, a plastic disassembly mat.

CHAPTER 8

SAFETY PRECAUTIONS

129. General

a. The safety of personnel is the primary concern of everyone during any exercise in which live ammunition is fired. To minimize the possibility of accidents, certain safety regulations have been established and currently are published in AR 385-63. The safety precautions given in this chapter are intended to emphasize and supplement the regulations in AR 385-63.

b. All officers conducting any type of firing are responsible for:

(1) Becoming familiar with all safety rules, general and local, pertaining to that type of firing and to the type weapons and ammunition being fired.

(2) Teaching all men on the ranges the meaning of these safety rules.

c. Each man is responsible for his own safety and that of others around him. Each man is required to enforce safety regulations on the range.

130. Safety Precautions, Mechanical Training

a. Start observing safety precautions as soon as you receive a weapon. Always check the chamber and receiver to insure that there is no live cartridge in the weapon.

b. Never playfully or carelessly point a weapon at anyone.

c. Check all dummy cartridges to make sure that there is no live ammunition among them.

131. Safety Precautions, Preparatory Marksmanship Training

a. Develop the habit of keeping the muzzle pointed up or down range.

b. Carry the weapon with the magazine removed, chamber empty, bolt forward, and cover closed.

c. Comply with all commands of the officer in charge and the coach.

132. Safety Precautions, Range Firing

a. Safety precautions *must* be observed on the range. An officer should be designated as safety officer, with the specific duty of insuring that all safety precautions are being observed. Each coach will require his firer to observe safety precautions.

b. Safety precautions listed above for mechanical training and preparatory marksmanship training also apply during range practice.

c. Precautions to be observed prior to firing include:
  (1) Do not draw or issue ammunition until the officer in charge of firing gives the command.
  (2) Do not start firing until:
    (*a*) All personnel are oriented regarding safety regulations and other pertinent information.
    (*b*) Roadblocks are established and road guards are posted.
    (*c*) A red streamer is displayed from a prominent place on the range.
    (*d*) All weapons are checked by an officer to insure that none contain live ammunition. The bores of weapons to be fired must be checked to insure that they contain no obstructions such as rust-preventive compound, cleaning patches, mud, or snow.
    (*e*) Markers are placed at the right and left safety limits, and personnel are warned not to fire outside of the markers.

d. Precautions to be observed during firing include:
  (1) Weapons will be loaded and unloaded on the firing line, on command of the officer in charge.
  (2) The commands COMMENCE FIRING and CEASE FIRING are given loud and clear (whistle signals may be used when necessary). Anyone who considers it necessary to insure safety may give the command CEASE FIRING. When a firer hears the command CEASE FIRING, he will take his finger off the trigger, close the cover, and wait for further instructions.
  (3) As soon as the firing exercise is completed, or on command, the firer will clear his weapon.
  (4) No one will move in front of the firing line unless directed to do so by the officer in charge, who, before giving this command, will have all weapons cleared, checked by the safety officer or coach, and grounded.
  (5) No weapon is moved in front of the firing line.
  (6) No weapon will be removed from the firing line until it has been checked, by the safety officer or one of his representatives, to see that it is clear.

(7) Proper care will be given ammunition (par. 53).

(8) All accidents will be immediately reported to the officer in charge. AR 385-63 prescribes the report that will be made for accidents involving faulty weapons or ammunition. SR 385-10-40 prescribes the report that will be made for accidents that are not the result of faulty weapons or ammunition.

(9) During moving target firing, no member of the pit detail will leave the pits until permission is granted by the safety officer.

(10) During the time that a member of the pit detail is out of the pits and in the target area, a red flag will be displayed at the pits.

(11) When guns are being fired from vehicles, a red flag will be displayed on each vehicle while the gun in that vehicle is loaded.

e. Precautions to be observed after firing include:

(1) All brass and live cartridges will be kept separate. Brass will be inspected to insure that there are no empty cartridge cases with unexploded primers.

(2) All cardboard ammunition cartons will be inspected to insure that they contain no live cartridges or brass.

(3) Before any weapon is removed from the firing line or from the range, it will be checked to insure that it contains no live ammunition.

(4) All personnel are inspected to insure that no unauthorized person is carrying live ammunition or brass from the range.

(5) All range flags are taken down, and the road guards are withdrawn from their posts.

APPENDIX

REFERENCES

AR 370-5	Qualification in Arms.
AR 385-63	Regulations for Firing Ammunition for Training, Target Practice and Combat.
SR 320-5-1	Dictionary of United States Army Terms.
SR 385-10-40	Accident Reporting.
FM 21-5	Military Training.
FM 21-6	Techniques of Military Instruction.
FM 21-30	Military Symbols.
FM 21-40	Defense Against CBR Attack.
TM 3-220	Decontamination.
TM 9-850	Abrasive, Cleaning, Preserving, Sealing, Adhesive, and Related Materials Issued for Ordnance Materiel.
TM 9-855	Targets, Target Material, and Training Course Layouts.
TM 9-2171-1	Cal .45, Submachine Guns M3 and M3A1.
TM 9-2205	Fundamentals of Small Arms.
TM 9-1990	Small Arms Ammunition.
DA Pam 108-1	Index of Army Motion Pictures, Filmstrips, Slides, and Phone-Recordings
DA Pam 320-1	Dictionary of United States Military Terms for Joint Usage.
DA Pam 310-3	Index of Training Publications.
DA Pam 310-4	Index of Technical Manuals, Technical Regulations, Technical Bulletins, Supply Bulletins, Lubrication Orders and Modification Work Orders.
DA Pam 310-5	Index of Graphic Training Aids and Devices.
ORD 7, SNL A-58	Submachine Gun Cal .45 M3 and M3A1.

INDEX

	Paragraph	Page
Accessories	48	40
Advice to instructors	109	83
After-firing care and cleaning	41	36
After-storage cleaning	44	38
Aids, training	126	90
Aiming. (*See* Sighting and aiming.)		
Allotment of hours:		
Familiarization course	113	84
Marksmanship course	112	83
Ammunition	49	40
Care, handling, and preservation	53	41
Classification	50	40
Destruction	108	82
Hangfires	56	43
Identification	52	41
Lot numbers	51	40
Precautions in firing	55	42
Range firing	122	87
Storage	54	42
Application of leads	90	73
Armorer, disassembly	8	11
Assembly:		
Field disassembly, after	12	19
Groups	14	22
Guides to follow	9	12
Housing, functioning	25	29
Assault position	72	53
Assistant instructors	110	83
Atomic attack, care and cleaning	45	38
Authority, destruction of materiel	104	82
Authorization for disassembly	8	11
Bar, sighting and aiming	70	48
Barrel, data	4	5
Before-firing care and cleaning	39	36
Biological attack, care and cleaning	45	38
Bolt and guide rod group:		
Assembly	14	22
Disassembly	13	19
Building ranges	85	71
Bursts, firing	73	54
Card, score	125	88
Care and cleaning	37	34
After firing	41	36
Ammunition	53	41
Before firing	39	36
CBR attack	45	38

	Paragraph	Page
Combat conditions	42	37
During firing	40	36
Preparation for storage	43	37
Unusual climatic conditions	46	39
Carrying position	58	44
Cartridge. (*See* Ammunition.)		
Causes of stoppages	35	33
Chambering	16, 20	25, 28
Characteristics of fire	101	80
Charts	128	91
Check, operation	15	24
Chemical attack, care and cleaning	45	38
Classification of ammunition	50	40
Cleaning. (*See also* Care and cleaning.)		
Materials	38	34
Weapons received from storage	44	38
Coaching	67	48
Moving target firing	95	75
Sighting and aiming exercises	70	48
Vehicular firing	100	77
Cocking	16, 24	25, 28
Cold climates, care and cleaning in	46	39
Collective firing	101	80
Commands, fire	102	80
Common stoppages	35	33
Construction:		
Ranges	85, 94, 98	71, 74, 77
Wooden working model	127	90
Courses:		
Allotment of hours	112–115, 118, 119	83, 86, 87
Moving target	95	75
Stationary target	77, 82, 83	58, 61, 65
Vehicular firing	99	77
Cycle of operation	16	25
Data	4	5
Decontamination	45	38
Description:		
Ranges	85, 94	71, 74
Submachine gun	3	2
Destruction of materiel	104	82
Detailed disassembly	13	19
Determination of leads	89	72
Differences in models	5	5
Disassembly	6, 8	7, 11
Field	10, 11	13
Groups	13	19
Guides to follow in	9	12
Dismounted range firing, procedure	82	61
Drivers for range firing	123	88
Dry climates, care and cleaning in	46	39
During-combat care and cleaning	42	37
During-firing care and cleaning	40	36

TAGO 7261-B, July

	Paragraph	Page
Ejection	16, 23	25, 28
Examination before range firing	75, 76	57
Exercises:		
Nonfiring marksmanship	74	56
Sighting and aiming	70	48
Technique of fire	103	80
Exposing targets	84	67
Extraction	16, 22	25, 28
Failures	33–35	32
Familiarization firing, procedure	80, 83	60, 65
Feeding	16, 19	25, 27
Field disassembly	10, 11	13
Filling the magazine	28	29
Fire commands	102	80
Firing	16, 21, 30	25, 28, 30
Familiarization	80, 83	60, 65
Instructor practice	78	59
Moving ground targets	87, 95	72, 75
Ranges. (See Ranges.)		
Record	79	60
Vehicles, from	97	76
Functioning	16, 17	25, 26
Housing assembly	25	29
Safety lock	26	29
Fundamentals of marksmanship	64	47
Gas attack, care and cleaning after	45	38
General data, description	3, 4	2, 5
Groups, disassembly of	13	19
Guide rod and bolt group:		
Assembly	14	22
Disassembly	13	19
Guides for disassembly, assembly	9	12
Handling of ammunition	53	41
Hangfires	56	43
Hot climates, care and cleaning in	46	39
Housing assembly:		
Disassembly	13	19
Functioning	25	29
Identification of ammunition	52	41
Immediate action	36	34
Important points, sighting and aiming	71	53
Importance of training	2	2
Individual:		
Score card	125	88
Soldier, disassembly by	8	11
Inspection:		
Arms	60	44
Ranges	124	88
Submachine gun	121	87
Instruction (see also Training)		
Practice firing	78	59
Instructors, advice	109	83

	Paragraph	Page
Kneeling position	72	53
Leads:		
Application	90	73
Determination	89	72
Use	88	72
Loading:		
Magazine	28	29
Submachine gun	29	29
Lock, safety, functioning	26	29
Lot number, ammunition	51	40
Lubrication	38–46	34
Magazine:		
Assembly	14	22
Disassembly	13	19
Loading	28	29
Maintenance. (See Care and cleaning.)		
Malfunctions	33, 34	32
Manipulation, trigger	73	54
Manual of arms	57	44
Carrying position	58	44
Inspection arms	60	44
Port arms	59	44
Present arms	61	45
Sling arms	62	45
Markings, ammunition	52	41
Marksmanship (see also Firing)		
Fundamentals	64	47
Training	63	47
Materials, cleaning	38	34
Mechanical training	114, 116, 118	84, 85, 86
Safety precautions	130	96
Method. (See Procedure.)		
Models:		
Differences in	5	5
Wooden working	127	90
Moving ground targets:		
Firing	87	72
Range	92	73
Nomenclature	7	7
Nonfiring mechanical exercises	74	56
Notes, training	116, 117, 120	85, 86, 87
Number, ammunition lot	51	40
Oil, lubricating	38	34
Operation (see also Functioning)		
Check	15	24
Filling magazine	28	29
Firing submachine gun	30	30
Loading submachine gun	29	29
Malfunctions	33, 34	32
Submachine gun	27	29
Unloading submachine gun	31	32
Ordnance personnel, disassembly	8	11

	Paragraph	Page
Parts:		
Nomenclature	7	7
Repair	47	40
Phases of training	65, 66	47
Port arms	59	44
Positions	72	53
Practice firing	78, 82	59, 61
Precautions. (*See* Safety precautions.)		
Preparatory marksmanship training	66	47
Notes	117	86
Schedule	115	85
Preparation of gun for storage	43	37
Present arms	61	45
Preservation of ammunition	53	41
Prevention of stoppages	35	33
Preventives, rust	38	34
Principles, destruction of materiel	105	82
Procedure:		
Assembly after field disassembly	12	19
Care and cleaning	38–46	34
Destruction of materiel	107, 108	82
Dismounted range firing	82	61
Familiarization firing	83	65
Field disassembly	11	13
Manual of arms	58–62	44
Moving target firing	95	75
Vehicular firing	100	77
Prone position	72	53
Qualification:		
Firing	79	60
Record	68	48
Questions, examination before firing	76	57
Radiological attack, care and cleaning after	45	38
Range:		
Dismounted practice	85	71
Firing, responsibility	81	60
Moving target	92	73
Safety precautions	86, 129, 132	71, 96, 97
Vehicular	98	77
Record:		
Firing	79, 82	60, 61
Instruction and qualification	68	48
Repair parts	47	40
Responsibility:		
Range firing	81	60
Safety precautions	129	96
Runaway gun	34	32
Rust preventives	38	34
Safety:		
Lock	26	29
Precautions	32	32
Ammunition	53, 55, 56	41, 42, 43

	Paragraph	Page
Mechanical training	130	96
Range	86, 96, 132	71, 76, 97
Responsibility for	129	96
Sample technique-of-fire exercises	103	80
Schedules, training	111	83
Score:		
Card	125	88
Qualification	79	60
Range firing	82, 83, 95	61, 65, 75
Vehicular firing	100	77
Sear and trigger group:		
Assembly	14	22
Disassembly	13	19
Functioning	18	26
Sequence of functioning	17	26
Sighting and aiming	69	48
Exercises	70	48
Important points about	71	53
Single shots, firing	73	54
Sitting position	72	53
Sling arms	62	45
Sluggish operation	34	32
Spare parts	47	40
Standing position	72	53
Steps in functioning	16	25
Stoppages	35, 82	33, 61
Storage:		
Ammunition	54	42
Cleaning weapons after	44	38
Preparation for	43	37
Table:		
Instruction practice	78	59
Familiarization firing	80	60
Leads	90	73
Record firing	78	59
Targets:		
Moving ground	93	74
Firing	87	72
Stationary, types	84	67
Technique:		
Fire	101	80
Moving targets	91	73
Trigger manipulation	73	54
Training:		
Aids	126	90
Destruction of materiel	106	82
Importance	2	2
Marksmanship	63	47
Phases	65	47
Preparatory	66	47
Notes	116, 117, 120	85, 86, 87
Record	68	48
Schedules	111, 114, 115, 118, 119	83, 84, 85, 86, 87

	Paragraph	Page
Trigger:		
Manipulation	73	54
Sear group. (*See* Sear.)		
Tropical climates, care and cleaning	46	39
Types:		
Ammunition	52	41
Fire commands	102	80
Uncontrolled fire	34	32
Unloading submachine gun	31	32
Use of leads	88	72
Vehicles:		
Firing from	97	76
Range firing	123	88
Wooden working model	127	90

[AG 472.5 (13 May 57)]

By Order of *Wilber M. Brucker,* Secretary of the Army:

MAXWELL D. TAYLOR,
General, United States Army,
Chief of Staff.

Official:
HERBERT M. JONES,
Major General, United States Army,
The Adjutant General.

Distribution:
Active Army:
Technical Staff, DA
Admin & Technical Staff Bd
Hq CONARC
OS Maj Comd
OS Base Comd
Log Comd
MDW
Armies
Corps
Div
Brig
Engr Gp
Inf Regt
Armor Gp
Engr Bn
FA Bn
Inf Bn
Ord Bn
QM Bn
Sig Bn
Armor Bn
MP Bn
AAA Bn
Cml Co
Engr Co
FA Btry
Inf Co
Ord Co
QM Co
Sig Co
Armor Co
MP Co
AAA Btry
Abn Co
USMA
Armor Sch
Inf Sch
Ord Sch
PMST Sr Div Units
PMST Jr Div Units
PMST Mil Sch Div Units
Mil Msn

NG: State AG; units—same as Active Army.
USAR: Same as Active Army.
For explanation of abbreviations used, see AR 320-50.

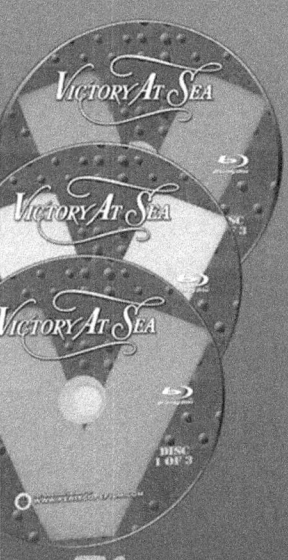

IN HIGH DEFINITION
NOW AVAILABLE!

COMPLETE LINE OF WWII AIRCRAFT FLIGHT MANUALS

WWW.PERISCOPEFILM.COM

©2013 Periscope Film LLC
All Rights Reserved
ISBN#978-1-940453-11-8
www.PeriscopeFilm.com

www.ingramcontent.com/pod-product-compliance
Lightning Source LLC
LaVergne TN
LVHW051845080426
835512LV00018B/3078